This book is an up-to-date text on electronic circuit design. The subject is dealt with from an experimental point of view, but this has not restricted the author to well-known or simple circuits. Indeed, some very recent and quite advanced circuit ideas are put forward for experimental work. Each chapter takes up a particular type of circuit, and then leads the reader on to gain an understanding of how these circuits work by proposing experimental circuits for the reader to build and make measurements on. This is the first book to take such an experimental approach to this level.

The book will be useful to final year undergraduates and postgraduates in electronics, practising engineers, and workers in all fields where electronic instrumentation is used and there is a need to understand electronics and the interface between the instrument and the user's own experimental system.

T0243065

Circuits for electronic instrumentation

Circuits for electronic instrumentation

T. H. O'DELL

Reader Emeritus, University of London

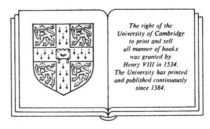

The right of the
University of Cambridge
to print and sell
all manner of books
was granted by
Henry VIII in 1534.
The University has printed
and published continuously
since 1584.

CAMBRIDGE UNIVERSITY PRESS

Cambridge

New York   Port Chester   Melbourne   Sydney

CAMBRIDGE UNIVERSITY PRESS
Cambridge, New York, Melbourne, Madrid, Cape Town, Singapore, São Paulo

Cambridge University Press
The Edinburgh Building, Cambridge CB2 2RU, UK

Published in the United States of America by Cambridge University Press, New York

www.cambridge.org
Information on this title: www.cambridge.org/9780521404280

First published 1991
This digitally printed first paperback version 2005

A catalogue record for this publication is available from the British Library

ISBN-13 978-0-521-40428-0 hardback
ISBN-10 0-521-40428-2 hardback

ISBN-13 978-0-521-01758-9 paperback
ISBN-10 0-521-01758-0 paperback

Contents

Contents

Preface

This book has been written for people who like to build electronic circuits and then experiment and make measurements on them. Such an experimental approach to understanding electronics does not mean that the work need be restricted to well-known or simple circuits. Some very recent circuit ideas are put forward as experimental exercises here, and the book should prove interesting both to students and to people, of all ages, who are already working in industry and research, and who would like to have experience of some recent developments in the field of electronic instrumentation.

In writing the book I have been greatly helped by conversation and correspondence with many people. In particular, Dr Asad Abidi, now at UCLA, Dr David Haigh, at UCL, London, and Dr Bhikhu Unvala, at ICSTM, London. Dr Unvala has also been most helpful in providing some of the experimental facilities. Finally, sincere thanks to Mr Ali Mehmed, whose careful reading of the typescript resulted in a number of changes for the better, along with many corrections.

London   T. H. O'Dell
July, 1990

1

Circuits for electronic instrumentation

1.1 Introduction

This book is concerned with the electronic circuits which are found in instrumentation systems. The interest here is with circuit detail. What might be called the 'difficult' circuits of instrumentation are those circuits which deal with the signals of interest *before* these are digitised. These are analog circuits, as a rule, and can involve problems associated with very high frequency technique, low noise level, d.c. stability, and so on.

Electronic instrumentation has developed greatly over the past half century. This development will be reviewed in the next section. Inside the instrument, however, at the level of circuit detail, it is interesting to see how often the same kind of circuit problem comes up again and again. This fact seems to suggest that circuit designers should take some interest in the history of their subject, and have a good knowledge of past technique as well as current technique. For this reason this book gives numerous references to the literature and, when possible, a note is given of the apparent origin of any widely used circuit technique.

The point about circuit detail, a particular circuit idea or what kind of circuit to use, also seems very important in electronic circuit design. It is impossible to calculate component values, device specifications or tolerances, before the circuit *shape* is fairly well determined. For example, in the trivial case of a single stage of amplification, is this to be grounded emitter, grounded base, or grounded collector? Is the stage to use a bipolar transistor anyway? Would a JFET be better? Why not choose an MOST? This initial step in the design process is often glossed over as being obvious or unimportant, but it is not. The really new ideas in electronics are often precisely within this field of circuit shape, and this concept is considered in detail in a separate section below.

Finally, one does not learn to be a circuit designer in the lecture theatre but in the laboratory. The most satisfying way to learn electronics is to build circuits and make measurements on these circuits. This may well mean restricting experimental work to the lower frequencies, simply because the most advanced constructional techniques and the latest test equipment may not be available to the student. Such restrictions should never deter the experimentalist in electronic circuits, and this book lays great emphasis on experimental work. Each chapter gives full details of experimental circuits which should be put together in order to see the full implications of the text.

1.2 Electronic instrumentation

The circuits which are discussed in the following chapters of this book have been taken from the field of electronic instrumentation. The main reason for this is that students should always have access to a circuits laboratory where test instruments are available, and they should be able to examine these closely. The handbooks of these instruments give the circuit detail of what might be called 'real' circuits, in contrast to the idealised circuits of the majority of text books, while the instruments themselves give examples of the hardware realisations of these circuits, probably using techniques which are more or less up to date.

A great deal can be learnt from this kind of close examination of the test equipment in a circuits laboratory. As mentioned above, however, the real business of the student of electronic circuit design is to build circuits and make measurements.

The development in electronic instrumentation over the past half century can be illustrated by choosing just one kind of instrument: the oscilloscope. In fact, the majority of the circuits in this book have been taken from this example, which covers a very wide field: the analog, sampling, and digitising forms of the oscilloscope.

The analog oscilloscope, in the form known today, was introduced in the early 1930s. The source of most of the ideas in this area, at that time, seems to have been Manfred von Ardenne [1], whose laboratory in Germany provided the cathode ray tubes (CRTs) for these first instruments, and also many of the circuit ideas. These early instruments gave bandwidths of only a few megahertz, used *RC* coupled amplifiers, and were difficult for one person to lift. Today, instruments with direct-coupled amplifiers, having millivolt sensitivity and bandwidths well over 100 MHz, are available, and weigh only a few kilograms. The deflection amplifier circuits and the waveform generating circuits of the analog oscilloscope have been chosen for study in chapters 6 and 7 of this book.

In the 1950s, an analog oscilloscope with a bandwidth of 30 MHz was considered to be a top of the line research instrument. This limited bandwidth was a severe restriction to engineers working in the rapidly developing field of digital electronics, where circuits were beginning to work at nanosecond speeds. There was also a great demand for a high speed oscilloscope from workers in the field of digital memories, and in nuclear physics. These pressures may have helped to bring about the development of the sampling oscilloscope.

The sampling oscilloscope uses an idea which is quite old. It is really like a stroboscope where a high speed repetitive event is viewed by means of a short flash of light, and this flash is made to occur at a slightly later time during each repetition. The technique was used by Lenz to view electrical waveforms, by taking a sample of the waveform with a contact that closed for a very short time, as early as 1849 [2].

The first commercial sampling oscilloscope appeared in 1950 [3] and had a bandwidth of 50 MHz. This was closely followed, in 1952, by a 300 MHz instrument [4]. The development of a solid state sampling oscilloscope by Chaplin, Owens and Cole [5], in 1959, marked the beginning of a radical change in technique, leading to the digitising oscilloscope of today.

The digitising oscilloscope, like the sampling oscilloscope, uses a sampling gate to take a measurement of the incoming signal over a very short time. This measurement is then put into the memory of the instrument, the next measurement is put into another memory location, and so on. In this way, a record of the incoming signal is built up, and this may be displayed later on, or as the data comes in, according to whatever program the user may decide. In what may be its most sophisticated form, the digitising oscilloscope is simply a small box with only coaxial sockets on the front, for Y inputs and trigger signals, and a connector on the back which links the digitising oscilloscope to a personal computer (PC) [6]. The VDU of the PC will then be a 'soft front panel' giving the oscilloscope display and all the 'controls', which are now accessed from the keyboard or from some other user interface. The most advanced digitising oscilloscopes, which may be able to accept data at up to one billion samples per second [7], are too fast to work directly with a PC and are stand-alone instruments.

Both the sampling oscilloscope and the digitising oscilloscope call for very short pulse generator circuits, and for high speed sample and hold circuits. These are, at first sight, very simple circuits which involve only a

few components. For this reason they have been chosen to be the subjects
of the next two chapters of this book. Chapter 4 deals with comparator
circuits, which are a key feature of the very fast analog to digital
converters (ADCs) found in the digitising oscilloscope. Chapter 5
considers circuits which are common to analog, sampling, and digitising
oscilloscopes: the probe and input circuits which are needed to connect
these instruments to the outside world.

The book closes with three chapters that consider circuit design
problems from other areas of instrumentation. Chapter 8 deals with
switched capacitor circuits, chapter 9 considers phase locked loop circuits,
and chapter 10 looks at the circuit techniques which are used to obtain low
noise. In every chapter there are experimental circuits which may be built.

1.3 Circuit shapes and circuit ideas

To return to the idea that was put forward at the beginning of this chapter,
there seem to be two quite separate steps in the circuit design process, as
there are in any design process. Designers first sketch out what kind of
circuit they plan to build, they sketch out its general shape, their circuit
ideas. No values, power supplies, device types, no *numbers*, are involved
at this stage. The *Gestalt* is the problem here, and this is why the term
circuit *shape* is chosen to express this stage in the design process. Only
when this circuit shape has been decided can any calculation of component
values, and then performance, be made. Certainly, the results of these
calculations may well cause the designer to go back and think about new
possibilities, new circuit shapes, but this does not alter the fact that the
first step in the design process is one of imagination and intuition. There
is no *algorithm* [8] which can be used to come up with a new circuit idea.

The process of invention, the way in which circuit designers arrive at
really new ideas, often seems to be a process of imaginatively combining
ideas that were new at some earlier time. Two examples of this will be
given in the next two sections. The first is from the late 1960s, when
electronic circuit design in monolithic silicon bipolar circuits was by far
the most active branch of the subject. The second example concerns
today's interest in gallium arsenide integrated circuits.

1.4 A new circuit shape in bipolar silicon

Gilbert has given a very interesting account of the way in which old and
well-established circuit ideas, which were, of course, once quite new and
even revolutionary, may be combined to provide a completely new circuit

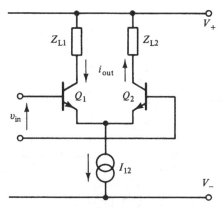

Fig. 1.1. *The long tailed pair.*

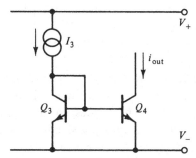

Fig. 1.2. *The Widlar current mirror.*

shape [9]. The process of invention which Gilbert describes is shown here in Figs 1.1–1.3.

In Fig. 1.1, the well-known long tailed pair circuit is shown. This is a very old circuit idea, first introduced by Blumlein [10] in 1936, which is now used for the input circuit of nearly all operational amplifiers. Attention is directed towards the output currents in Fig. 1.1, and the collector loads are represented by quite general impedances. As it stands, the long tailed pair provides a very non-linear and temperature dependent transconductance, di_{out}/dv_{in}, and this was one property which Gilbert [9] aimed to improve.

Fig. 1.2 shows another very well-known circuit shape: the Widlar current mirror. Again, this is an old circuit idea, which originated in 1965 [11]. In this circuit, Q_3 is clamped active by the simple expedient of connecting its base to its collector. Because the emitter junctions of Q_3 and Q_4 are in parallel, both devices must have the same V_{BE}, so that, provided they are identical devices and are both at the same temperature, as they would be in one and the same integrated circuit, i_{out} must equal I_3. The

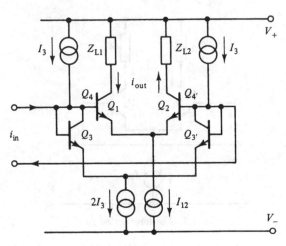

Fig. 1.3. *Gilbert's 'marriage' of Figs.* 1.1 *and* 1.2.

collector of Q_4 may be connected to any potential above V_{BE}, always bearing $V_{CE(max)}$ in mind.

Fig. 1.3 shows what Gilbert [9] called the 'marriage' of Figs. 1.1 and 1.2. The circuit shown in Fig. 1.2 is introduced into the circuit shown in Fig. 1.1 as an input circuit, on both sides, so that the new circuit, shown in Fig. 1.3, accepts a current as its input signal. In fact, the current gain, i_{out}/i_{in}, is given by I_{12}/I_3 [12], and is linear over the range $-I_3 < i_{in} < +I_3$ as well as being temperature independent. The gain–bandwidth product of the new circuit is essentially equal to that of the transistors used, and Gilbert's paper [9] described an integrated circuit using five such stages of amplification.

Gilbert's circuit has not been widely adopted as a wide-band current amplifier [13], and this is precisely the reason it has been chosen here as an example of a new circuit shape. Although the circuit uses the well-established bipolar technology, and is over 20 years old, it may well be unfamiliar to many readers, whereas the circuits shown in Figs. 1.1 and 1.2 will be absolutely familiar and are circuit shapes which are accepted today without question. This was certainly not the case when these circuits were first published, however.

To find a second example of a new circuit shape, which may strike many readers as a new and interesting circuit idea, it is only necessary to look at the kind of circuit design work that is taking place today in the field of gallium arsenide integrated circuits.

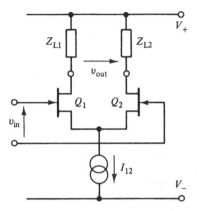

Fig. 1.4. *The long tailed pair in gallium arsenide.*

1.5 A new circuit shape in gallium arsenide

Gallium arsenide integrated circuits make use of n-channel field effect transistors (FETs). The gates of these devices are Schottky barriers, hence the use of the usual JFET symbol, and the devices have excellent high frequency properties due to the high electron mobility in gallium arsenide, as well as the possibility of making FETs with very short channels.

Fig. 1.4 shows the long tailed pair circuit again, this time drawn using a pair of gallium arsenide FETs. The intention of the circuit designer is to make a high gain voltage amplifier from this well-known circuit shape. This is not easy in gallium arsenide because the FETs have modest transconductance, dI_d/dV_{gs}, coupled with a rather high output conductance, dI_d/dV_{ds}. The voltage gain, of course, can never exceed $(dI_d/dV_{gs})/(dI_d/dV_{ds})$, no matter what values are chosen for Z_{L1} and Z_{L2}.

Fig. 1.5 shows another classical electronic circuit, the cascode [14], now drawn in gallium arsenide technology. This well-known amplifier circuit provides a high input impedance, at high frequency, compared to a single grounded source stage, because Q_3 has its gate connected to a constant bias voltage, V_B, and thus holds the drain of Q_4 at a fairly constant level. The output signal appears at the drain of Q_3, and this means that the feedback capacitance from output to input, in Fig. 1.5, should be very small.

The cascode has been the subject of a paper by Abidi [15], who shows that there is a great deal more involved in understanding this circuit than is generally thought. In this paper [15], and in an earlier paper [16], Abidi discusses the circuit shown in Fig. 1.6. This circuit has become widely used in gallium arsenide integrated circuits, and it is possible to argue that its origin is a combination of the two classical circuits shown in Figs. 1.4 and

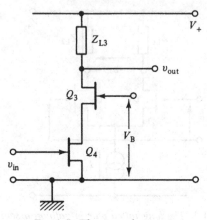

Fig. 1.5. *The cascode circuit.*

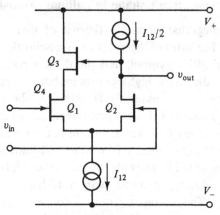

Fig. 1.6. *The gain enhanced circuit in gallium arsenide, formed by the combination of Figs.* 1.4 *and* 1.5.

1.5. As far as the two circuit shapes are concerned, this argument seems sensible. However, some very interesting changes in the function of the two transistors, Q_3 and Q_4, which came into the new circuit from the cascode circuit shape, may be seen. In Fig. 1.6, Q_3 is now driving the drain of Q_4, which is, of course, the Q_1 of the original long tailed pair shown in Fig. 1.4. The drain of this Q_1 is now 'bootstrapped' [16], in that its voltage rises as v_{out} rises, whereas, in the original long tailed pair circuit, Fig. 1.4, the drain of Q_1 falls when the drain of Q_2 rises.

Some kind of d.c. feedback must be arranged across the new circuit, shown in Fig. 1.6, in order to define the d.c. level at the drain of Q_2. The final result then *is* a high gain voltage amplifier with a gain of the order of the *square* of the gain of the simple circuit shown in Fig. 1.4, and this may be taken as a good example of invention in the field of electronic circuit design. A similar gain enhancement idea, coming from the same combination of the cascode and long tailed pair circuits, is found in CMOS integrated operational amplifier design [17].

1.6 Method in design

The circuits which feature in the following chapters of this book are presented in the light of the above discussion. For example, in chapters 2 and 3, where the circuits are fairly simple, the search for a good circuit shape, a good circuit idea, appears to involve a choice of a circuit from the past literature. In chapter 4, where the fast comparator circuits of the flash ADC are considered, the problem of circuit shape should come over more in the spirit of Figs. 1.1–1.6. The comparator circuit finally chosen for experimental work is a combination of two long tailed pairs, the circuit shown in both Figs. 1.1 and 1.4. The same story appears in chapter 6, where series and shunt feedback circuits are combined to give wide-band amplifier circuits; and in chapter 8, where a voltage controlled oscillator (VCO) circuit from chapter 7 is combined with a simple charge pump to give an ultra-linear VCO.

When such a variety of circuit shapes is presented, the question of method in design should be brought up again. Is there, in fact, a method which is followed when a designer comes up with a new circuit shape, like the circuits shown in Figs. 1.3 and 1.6, or must this be accepted as an intuitive or imaginative process, as it was above in sections 1.3–1.5? The answer to this important question is obscured by the fact that, once the new circuit shape is known, it *is* possible to give an account of the reasoning that might have led to this new idea. This is done in many of the examples which are given in the following chapters.

Finally, it must be remembered how the new and unexpected can turn up regularly, in any field of technology, and make the technique of the recent past look quite outmoded. In electronics there are many examples: the transistor of the mid-1950s, the silicon integrated circuit of the mid-1960s, the VLSI circuits, particularly the microprocessor, of the mid-1970s, and recently the gallium arsenide integrated circuits of the mid-1980s. All these revolutions have caused people to think about circuits in a new way, to think of quite new circuit shapes, as sections 1.4 and 1.5 illustrated. These same sections, however, showed that circuit ideas from earlier times were still influencing the process of invention.

1.7 Experimental circuits

The experimental circuits, which are given in the following chapters, have been chosen so that they may be put together in the simplest kind of circuits laboratory. Very fast, and very high frequency, circuits are slowed down to make construction quite straight-forward. This is also necessary because it is assumed that readers may only have fairly modest test equipment available.

In cases where the circuit interest is in the detail of, what would be, a large scale integrated circuit, as it is in chapters 4 and 9, it is possible to do very instructive experimental work by using transistor arrays, or by using an example of the integrated circuit itself which happens to have a pin-out that provides access to the signals needed.

The constructional methods which are used for experimental work are left fairly undefined in most cases because these must depend upon what the reader has available. One general rule which must be followed is that the layout of the experimental circuit must be thought about very carefully before construction begins. In this connection, the fact that these circuits are intended for experimental work, which means access for measurement, as well as a need to add or change a few components, must be borne in mind. Horowitz and Hill have a useful chapter on constructional techniques in their book [18], but their advice is directed more towards circuits that are built with the confidence that these circuits will form part of a finished product.

1.8 Symbols and abbreviations

Finally, a note on the symbols and abbreviations which will be used in the following chapters.

Voltages and currents are represented by upper case letters (V and I) when these are constant levels. The same rule applies for subscripts, so that V_{BE} is the bias voltage across a base-emitter junction, V_{DS} the voltage from drain to source, and so on. Small signals, which are superimposed upon such constant levels, appear in the text as lower case letters, both v and i, and the subscripts, v_{be} and v_{ds} for example. The use of lower case subscripts with upper case letters means a level which is changed by the experimentalist, for example, V_c would be a control voltage, I_{in} would be a direct current that might be varied from run to run, and so on.

When circuits involve many components, symbols like I_{C3} appear for the collector current of transistor Q_3 and, less elegantly, I_{R_3} for the current in R_3. With these conventions in mind, the reader should find the symbolism straightforward. All other symbols are defined when they first appear.

Abbreviations, like ADC, PLL, and so on, are defined when they first appear. The only one which deserves special comment is the troublesome 'd.c. level'. In this text, d.c. means direct current and d-c means direct-coupled. The term d.c. level is so widely used to mean the constant voltage level, or the mean voltage level, which appears at some point in a circuit, that to avoid its use seems to be pedantic.

Notes

1 von Ardenne, M., *Ein glückliches Leben für Technik und Forschung*, Kindler Verlag, Zurich and Munich, 1972.
2 Laws, F. A., *Electrical Measurements*, McGraw-Hill, New York, 1917, p. 613.
3 Janssen, J. M. L., *Philips Tech. Rev.*, **12**, 52–59 and 73–82, 1950.
4 McQueen, J. G., *Electronic Engineering*, **24**, 436–41, 1952.
5 Chaplin, G. B. B., Owens, A. R., and Cole, A. J., *Proc. IEE*, **106B**, Suppls. 15–18, 815–23, May 1959.
6 An entire issue of *Energy and Automation* (Vol. X, April 1988) was devoted to the soft front panel PC-based instruments that were introduced by Siemens AG in 1987. The *Hewlett–Packard J*, **37**, No. 5, May 1986, was similarly devoted to a description of the Hewlett–Packard range of PC-based instruments, but these were withdrawn in 1989.
7 Millard, J. K., *Hewlett–Packard J.*, **39**, No. 3, 58–9, June 1988.
8 An algorithm is a list of instructions which can be followed by a person (or, of course, by a machine), who has no interest in, or understanding of, the problem the algorithm is intended to solve. A book which supports the idea that algorithms cannot be found which can reproduce 'all mental qualities – thinking, feeling, intelligence, understanding, consciousness –', to quote from p. 17 of the book, is R. Penrose's, *The Emperor's New Mind*, Oxford University Press, Oxford, 1989.
9 Gilbert, B., *IEEE J. Sol. St. Circ.*, **SC-3**, 353–65, 1968.
10 Blumlein, A. D., British Pat. No. 482740, 1936.
11 US Patent No. 3364434, to R. J. Widlar and the Fairchild Corp.
12 O'Dell, T. H., *Electronic Circuit Design*, Cambridge University Press, Cambridge, 1988, pp. 116–17.
13 Despite this, Gilbert's paper (note [9] above) was cited over 60 times in the *Science Citation Index* between 1968 and 1988.
14 The circuit may have been first introduced by F. V. Hunt and R. W. Hickman, *Rev. Sci. Inst.*, **10**, 6–21, 1939.
15 Abidi, A. A., *IEEE J. Sol. St. Circ.*, **SC-23**, 1434–7, 1988.
16 Abidi, A. A., *IEEE J. Sol. St. Circ.*, **SC-22**, 1200–4, 1987.
17 Gray, P. R., and Meyer R. G., *Analysis and Design of Analog Integrated Circuits*, John Wiley, New York, second edition, 1984, pp. 752–5.
18 Horowitz, P., and Hill, W., *The Art of Electronics*, Cambridge University Press, Cambridge, second edition, 1989, pp. 827–61.

2

Sampling pulse generator circuits

2.1 Pulse width and rise time

The problems considered in this chapter concern devices and circuits which can produce very short pulses. The sampling pulse generators, found in the most advanced sampling and digitising oscilloscopes, may produce pulses as short as 50 ps. The first point to consider, however, is the shape of the pulse itself. This might seem to be a very trivial matter, but it does hold the key to good pulse generator circuit design.

Fig. 2.1 illustrates what is meant. Every real pulse has finite rise and fall times, τ_r and τ_f in Fig. 2.1, as well as a pulse width, τ. The rise and fall times are traditionally taken between the times at which the level is at 10% and 90% of the pulse amplitude, A. For the sampling pulse generator, the pulse width, τ, is best taken as shown in Fig. 2.1: the time that the pulse lies above 90% of its maximum amplitude.

What is important here is that a well-designed pulse generator circuit will always have some clearly identifiable part that defines the pulse width. For example, this could be a length of transmission line or a CR time constant. Whatever defines the pulse width, it must have negligible effect upon the rise and fall times, and these both need to be made as short as possible. In the case of a sampling pulse generator, where a very small pulse width is of primary concern, it is vital to obtain very short rise and fall times.

2.2 Circuit shape

In chapter 1, great emphasis was put upon the concept of circuit shape: the form that a circuit diagram has as a preliminary design sketch, the first idea the designer has, the circuit as it is before any calculations of circuit values can be made.

12

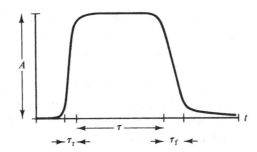

Fig. 2.1. *Defining the rise time, fall time, and width of a pulse.*

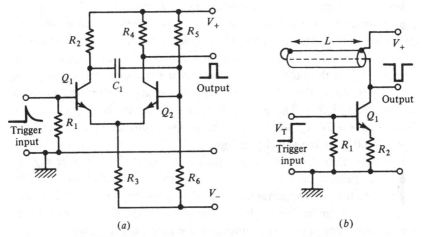

(a)   (b)

Fig. 2.2. *Two contrasting pulse generator circuit shapes.*

Fig. 2.2 shows two pulse generator circuit shapes to illustrate this point of view. In Fig. 2.2(a), the well-known monostable multivibrator [1] is shown. This is a very old circuit idea, and may have been first described by White [2]. In the form shown here, transistor Q_2 is normally held on, this being arranged by the choice of the bias resistors R_5 and R_6. This means that Q_1 is held off because of the emitter coupling between Q_1 and Q_2. A short trigger input pulse drives Q_1 on and Q_2 off. Q_2 remains off while C_1 charges up, and the time constant that determines how long Q_2 stays off, assuming that the devices do not saturate, is

$$T = C_1[R_2 + R_5 R_6/(R_5 + R_6)]. \qquad (2.1)$$

Unfortunately, this time constant must also determine how rapidly the emitter junction of Q_2 begins to turn on again at the end of the pulse, and this means that the fall time of the output pulse must depend to a certain extent upon the pulse length. This contradicts the expectations of the final paragraph of section 2.1: whatever decides pulse width should not effect rise or fall times.

Fig. 2.2(*b*) shows another possible circuit shape for a pulse generator. This is also a very old circuit idea, patented by Blumlein in 1940 [3]. The circuit shown here involves only a single transistor and a length of transmission line which has a good short circuit made at its far end. The transistor is normally off until an input step, V_T, turns Q_1 on so that it takes a current $I_C = (V_T - V_{BE})/R_2$. The transistor does not saturate.

The voltage level at the collector of Q_1 will drop by an amount $I_C Z_o$, where Z_o is the characteristic impedance of the transmission line. If this is a good quality low loss line, Z_o will be almost a pure resistance and the initial edge of the output pulse, which is negative going in this circuit, is determined mainly by the transistor turn-on time. The same applies to the trailing edge of the pulse because, again if the transmission line is good, this edge is simply the reflection of the initial edge, reflected and inverted by the short circuit. The pulse length in this circuit is very accurately defined as $\tau = 2L/v_p$, where v_p is the propagation velocity of the line: typically 60% of the velocity of light.

It follows that the circuit shape shown in Fig. 2.2(*b*) might be a good choice for generating the very short pulses that are needed for sampling circuits. As the pulse length is determined by a length, *L* in Fig. 2.2(*b*), and light travels 30 cm in 1 ns, the lengths involved for the generation of pulses considerably less than 1 ns in duration are conveniently small.

The problem which is obviously outstanding in the circuit shown in Fig. 2.2(*b*) is the switching device. No bipolar transistor can be expected to turn on in a time which is very much smaller than 100 ps. It is necessary to find some other kind of switching device.

2.3 The step recovery diode

The switching device that is almost universally used in sampling pulse generator circuits is the step recovery diode. This was originally called the charge storage diode, in an important paper dealing with the theory of this new device [4]. The terms snap-off diode and snap-back diode are also used.

The step recovery diode represents a rather unexpected change in the direction of fast switching device development. Earlier work had concentrated upon two main lines, both using three terminal active

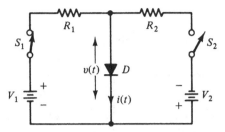

Fig. 2.3. *An experimental set-up for the study of reverse recovery in a diode.*

devices. These two main lines are conveniently illustrated using Figs. 2.2(*a*) and 2.2(*b*) again.

One line of development attempted to produce very fast switching by using amplifying devices, of very large gain–bandwidth product, in regenerative switching circuits of the kind shown in Fig. 2.2(*a*). Such circuits can provide very fast turn-on, much faster than the rise time of the input pulse that triggers this turn-on. However, the limitation is in the bandwidth of the circuit, and it is very difficult to generate well-shaped pulses, in the sense of Fig. 2.1, which are much shorter than 1 ns with circuits using the regenerative idea.

A second line of development involved circuits of the shape shown in Fig. 2.2(*b*), but using some kind of three terminal triggerable device instead of the simple amplifying device shown here. Such devices as thyratrons, triggered spark gaps, mercury wetted relays, avalanche transistors, four layer solid state devices, and an interesting cold cathode gas filled device known as a krytron, can all be found in the extensive literature on this topic. These triggerable devices do have very high speed possibilities, but they all suffer from jitter: there is a considerable statistical fluctuation in the time between the arrival of the trigger input pulse and the actual turn-on of the device.

The step recovery diode is a radical break in the tradition of fast switching. In the first place, it is a *two* terminal device. Secondly, its fast properties are associated with the way in which it turns *off*.

2.4 Step recovery diode theory

A step recovery diode is usually a silicon diode which has been made in such a way that it shows two special features in the reverse recovery mode: long minority carrier storage time followed by very short turn-off time. Fig. 2.3 shows a simple experimental set-up for the study of reverse recovery which will make clear what is meant here.

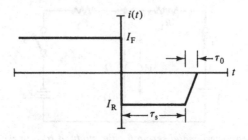

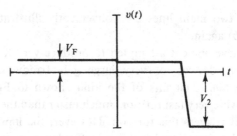

Fig. 2.4. *The waveforms that would be observed from the circuit shown in Fig. 2.3.*

In Fig. 2.3, a silicon diode, D, is shown connected to two voltage sources, V_1 and V_2, via current limiting resistors R_1 and R_2, and with switches, S_1 and S_2, to control the experiment. This is, of course, a thought experiment at this stage. An experimental circuit will be given towards the end of this chapter so that the reader can make real measurements.

As Fig. 2.4 shows, for $t < 0$, when S_1 is closed and S_2 is open, the diode takes a forward current I_F and has a forward voltage V_F across it. In the case of silicon, V_F is close to 0.7 V for a very wide range of current.

At $t = 0$, S_1 is opened and S_2 is closed. This causes a reverse current to flow through the diode, and Fig. 2.4 shows that the values of V_2 and R_2, in Fig. 2.3, have been chosen, in this thought experiment, so that $I_R = I_F$.

The reversal of current through the diode has very little effect upon the voltage across it: this is still close to 0.7 V. The reason for this is that a silicon pn junction stores charge when it is put into forward conduction, and it is not possible to change the voltage across the junction until there is a considerable change in the minority carrier density close to the junction region [5, 6, 7]. The small drop in forward voltage, shown in Fig. 2.4, is due to the reversal of the ohmic drop across the bulk of the diode structure.

It follows that quite a large reverse current can flow for a storage time, τ_s, as shown in Fig. 2.4, and then the diode does really turn off. The

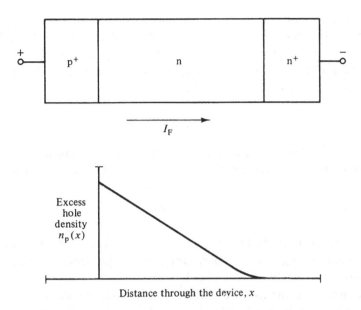

Fig. 2.5. *A diode with a p^+nn^+ structure in forward conduction and the resulting minority carrier density across the n region.*

current falls to zero and the voltage across the diode becomes V_2, in the reverse direction, and all this happens in turn-off time τ_0. The actual values of τ_s and τ_0 are determined by doping levels and geometry, and these are considered next.

In an ordinary, small signal, silicon diode, like the classical 1N914 device, every attempt is made by the device designer to reduce the total reverse recovery time, $\tau_s + \tau_0$ in Fig. 2.4. Designers of step recovery diodes aim for quite different behaviour. It will be seen in section 2.7 that circuit design requirements call for a storage time of a few nanoseconds, for a step recovery diode that is to be useful in a sampling pulse generator circuit. The turn-off time, τ_0, in contrast, must be made as short as possible. This can be achieved by using a diode geometry of the kind shown in Fig. 2.5: a p^+nn^+ structure.

In Fig. 2.5, the p^+nn^+ diode is shown passing a forward current, I_F. The p^+ and n^+ regions are very heavily doped, while the n region is very lightly doped and is thin compared to a diffusion length. It follows that the minority carrier density in the n^+ and p^+ regions can be ignored in order to obtain a qualitative picture of what is happening. What is important is the density of holes in the n region, and this is shown in Fig. 2.5. At the p^+n junction, which is abrupt, there is a high density of holes injected from the p^+ region. The density of minority carriers falls almost linearly across the thin n region, to become very small at the nn^+ junction. This happens because of the two boundary conditions at the nn^+ junction: (1) continuity of the slope of $n_p(x)$, and (2) $n_p(x) \approx 0$. The first condition gives continuity to the hole current density,

$$J_p = -eD_p\, dn_p(x)/dx \tag{2.2}$$

where e is the electronic charge and D_p is the diffusion constant for holes. The second condition comes from the fact that the diffusion length for holes,

$$L_p = (D_p\, T_p)^{\frac{1}{2}} \tag{2.3}$$

will be very small in the heavily doped n^+ region where the hole lifetime, T_p, is small and $n_p(x)$ must vanish rapidly as x continues to increase. D_p depends only slightly upon doping level.

It follows that the forward current in this diode is mainly carried by majority carriers in the p^+ and n^+ regions, but in the n region the forward current is a diffusion current of minority carriers, holes, and this diffusion current is down the gradient shown in Fig. 2.5. When the current through the diode is reversed, using the thought experiment shown in Fig. 2.3, it is not possible for the excess hole density distribution, shown in Fig. 2.5, to change very rapidly.

What will happen is shown in Fig. 2.6. Just after the reversal of the current, the excess hole density must drop within the part of the n region which lies close to the p^+n junction. This reverses the slope of $n_p(x)$ and causes a diffusion current in the direction of I_R. Note that a diffusion current must still flow in the old direction, the direction of I_F, over the remaining part of the n region, where the slope of $n_p(x)$ remains, for the moment, unchanged. The new diffusion current, near the p^+n junction, is greater, however.

Fig. 2.6 shows the continuation of the process as $n_p(x)$ reverses slope over more and more of the n region, the maximum value of $n_p(x)$ moving away from the p^+n junction and the amplitude of $n_p(x)$ falling as time passes. The key to the very fast turn off in the step recovery diode can be seen in the last curve: the one for $t = \tau_s$. This shows that nearly all the excess minority carrier charge has vanished at the same instant that $n_p = 0$ at the p^+n junction. It is only when $n_p = 0$, at the p^+n junction, that the reverse voltage across the junction can develop. If there are still some minority carriers left within the n region, these will flow across the p^+n junction and slow up this establishment of reverse voltage.

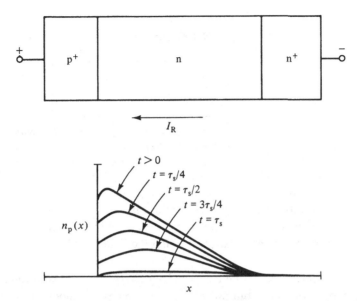

Fig. 2.6. *The excess hole density as a function of space and time during the flow of reverse current, I_R, in the p^+nn^+ diode. Note that the voltage across the diode is still in the forward direction.*

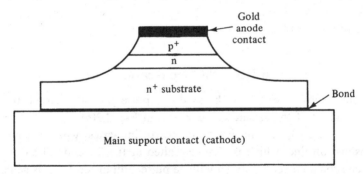

Fig. 2.7. *The mesa structure which is used for the step recovery diode.*

2.5 Step recovery diode design

A real step recovery diode, in contrast to the academic representation shown in Figs 2.5 and 2.6, will take the form shown in Fig. 2.7 [8]. This is the so-called 'mesa' structure, a Spanish word which first came into the English language as a technical term in physical geography to describe a high table land which is a remnant of a former plateau. On a microscopic scale, this is exactly what the mesa diode, shown in Fig. 2.7, is.

The first step in the fabrication process is to diffuse a structure, of the kind shown in Fig. 2.8, into an n^+ substrate. A gold contact is then

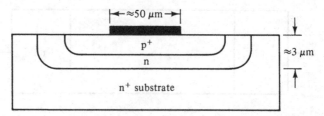

Fig. 2.8. *The step recovery diode structure before plasma etching. Note the different scales horizontally and vertically. Also the thickness of the n^+ substrate, as in Fig. 2.7, has been considerably reduced for this diagram.*

evaporated onto the centre of the p^+ region. By means of plasma etching, all the silicon which is not covered by the gold top contact is then removed, down to the n^+ substrate, leaving the structure shown in Fig. 2.7 behind.

The p^+nn^+ structure, which has been brought out in the discussion here, is the most common, but an n^+pp^+ step recovery diode may have advantages in speed [9]. The reason for using the mesa structure is that this gives a device which can be put into a package having negligible self capacitance. A step recovery diode is usually found in a minute ceramic package suitable for surface mounting on a ceramic substrate as part of a thick film circuit.

2.6 Using the step recovery diode

The step recovery diode can be used in a pulse generator circuit that is a development of the circuit shape shown in Fig. 2.2(*b*).

In Fig. 2.2(*b*), a transistor is shown in series with a length of transmission line, which is short circuited at the far end. The transistor switches on a current I_c and a voltage pulse, amplitude $I_c Z_o$, is generated across the transmission line.

The step recovery diode can only switch a current *off*. This means that, to produce a pulse, the device must be put in *parallel* with a length of transmission line and also in parallel with a constant current source, I. When the diode turns off, the current I will be switched into the line.

Now this is all very well, but it is obvious that this new parallel circuit cannot use a simple short circuit at the far end of the line. What is needed in this case is a large capacitor, in series with the input to the short circuited line, which will allow the d.c. level across the diode, V_F in Fig. 2.4, to become established, and yet present a negligible impedance to all the high frequency components of the very short pulse.

This is the shape of the experimental circuit which is described in the next section.

2.7 An experimental circuit

It was stated, at the beginning of this chapter, that a sampling pulse generator would be designed to produce a very short pulse, even as short as 50 ps. Circuits which work at such high speeds are usually constructed as hybrid circuits, using very small surface mounted components on a ceramic substrate with thick film printed interconnections.

The majority of readers may not have the facilities to make hybrid electronic circuits, using real step recovery diodes, or the very high speed test equipment which is needed to examine the dynamic behaviour of real step recovery diode circuits. For these reasons, an experimental circuit is described in this section, which is identical, as a circuit shape, to a real step recovery diode sampling pulse generator circuit, but which works at a much slower speed and can, therefore, be made up using quite conventional techniques.

Becker and Fomm [10] pointed out some years ago that many diodes which are designed as tuning diodes, varactor diodes, or what are often termed 'varicap diodes', can be used as step recovery diodes working at nanosecond switching speeds. This is to be expected because the tuning diode has the same p^+nn^+ or n^+pp^+ structure of the step recovery diode, but it is normally used under conditions of reverse bias and the reason for the very lightly doped n or p layer, in the tuning diode, is to obtain a wide variation in the capacitance of the reverse biassed p^+n, or n^+p, junction.

Tuning diodes are very easily available, in quite conventional types of package, and at low cost. This is because these diodes are widely used in television and FM tuners.

Fig. 2.9 shows the experimental circuit. Transistor Q_1 is normally off, and the step recovery diode, D_2, is in forward conduction because Q_2 is connected as a simple constant current sink. R_3, R_4 and R_6 are chosen to make this forward current in D_2 a few tens of milliamperes when V_- is at -15 V, but the actual value can be varied by using a variable negative supply. Q_2 is a simple general purpose transistor.

Q_1 is also a general purpose transistor, able to turn on a current of about 200 mA, maximum, when a negative going pulse, 5 V in amplitude, arrives at the input socket. R_1 and R_2 are chosen to give a 50 Ω termination to the input cable, C_1 being large, and the diode D_1 clamps the pulse input, from the point of view of Q_1, to the positive rail. (The idea of diode clamping is explained in all the elementary electronics texts [11], and will turn up repeatedly in the experimental circuits described in

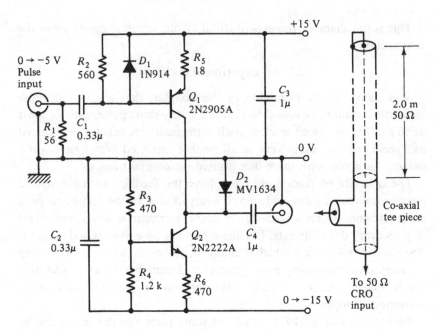

Fig. 2.9. *An experimental pulse generator circuit using a tuning diode,*
type MV1634, to play the part of the step recovery diode, D_2.

this book.) R_5 is then chosen so that the current turned on by Q_1,
$(V_T - V_{BE})/R_5$, V_T being the amplitude of the input pulse, is 200 mA
when V_T is 5 V.

Now the current that is turned on in Q_1 flows into the constant current
sink, Q_2, and, because it exceeds this current, a reverse current, $(I_{C1} - I_{C2})$,
is established in the diode D_2. However, because Q_1 is an ordinary general
purpose device, this current takes, perhaps, 10 ns to become established,
and it should now be clear why a step recovery diode is designed to have
a storage time of a few nanoseconds: it is necessary to establish the reverse
current well in advance of the fast recovery.

Referring to Fig. 2.9, it can now be seen how this circuit generates a
very short and well-shaped output pulse. When D_2 does turn off, or snap
off, to use a more emphatic description, the current $(I_{C1} - I_{C2})$ is diverted,
via C_4, into the two 50 Ω transmission lines which are connected in
parallel to the co-axial socket shown on the right-hand side of Fig. 2.9. A
voltage, $25(I_{C1} - I_{C2})$ V, is thus established across these transmission lines,
but can only persist for the time taken for this voltage step to propagate
along the 2 m length of line, to the short circuit, and then back again to
collapse the voltage at the tee piece. As this total round trip of 4 m, at a
velocity of, perhaps, 2×10^8 m/s, will take 20 ns, this should be the

duration of the output pulse which would be seen on the oscilloscope connected to the far end of the other line, shown on the lower right-hand side of Fig. 2.9.

2.8 Constructional work

The circuit shown in Fig. 2.9 may be built up in a very conventional way, using a small circuit board, provided D_2 and C_4 are connected very close to the co-axial socket which is shown on the right-hand side of Fig. 2.9. The point is to keep the inductance of this part of the circuit as low as possible, so that very short leads are called for as well as a careful choice for C_4. A low inductance capacitor can be built up by putting several small ceramic capacitors in parallel.

The short circuit at the far end of the 2 m length of cable must also be of very low inductance. The simplest way of doing this is to slide back the braid of the cable, strip off the inner conductor's insulation, and then pull the braid over the inner conductor for a few millimetres and solder through the short circuit this forms. The kind of cable used is also important, but this point is raised in the next section.

2.9 Experimental work

Fig. 2.10 shows an oscillograph of what was actually observed with the circuit shown in Fig. 2.9, using the values and devices given on the figure. The MV1634 tuning diode which was used does not show a very fast step recovery, but it does give a well-shaped pulse when $I_{c1} = 180$ mA and $I_{c2} = 20$ mA. This means $I_{c1} - I_{c2} = 160$ mA and the output pulse is 4 V in amplitude. The positive power supply, V_+, was 15 V, while V_-, and the amplitude of the negative input pulse, were both adjusted to optimise the pulse shape. The delay in the diode turning off can be increased by increasing the magnitude of V_-, as this increases the diode forward current, I_{c2}. Similarly, increasing the amplitude of the input pulse will reduce the delay because the net reverse current, $(I_{c1} - I_{c2})$, is increased. Both adjustments influence the pulse shape because the ideal situation, shown in Fig. 2.6, where the stored charge vanishes at the same instant $n_p(x) = 0$ at the junction, is more, or less, approached.

Other tuning diodes can be found which behave in a very similar way to the MV1634. The BA138 diode gives a very similar shape of output pulse, of smaller amplitude because it has to operate at a larger value of I_{c2} in order to delay step recovery until Q_1 has been turned on fully. The BA142 diode has a faster step recovery than the MV1634, and will begin to show the limitations of a 250 MHz oscilloscope. A real step recovery

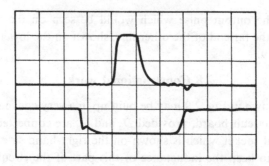

Fig. 2.10. Waveforms observed from the experimental circuit of Fig. 2.9. The time base is 20 ns/div. The lower waveform is the input pulse on 5 V/div. The upper waveform is the output pulse on 2 V/div.

diode, the 5082–0180 from Hewlett–Packard, is available in a conventional package but its recovery time is much too short for any analog oscilloscope to display it, and it is an expensive device: a selection of some 20 tuning diodes may be purchased for the price of a single 5082–0180.

The negative going edge of the output pulse, shown in Fig. 2.10, is slower than the rising edge. Referring to Fig. 2.1, this means that $\tau_f > \tau_r$, and there are two reasons for this. The first is that the negative going edge is only a reflection of the rising edge up to the time it reaches the co-axial tee piece, shown in Fig. 2.9. It then has to cross this tee piece and suffer the distortion due to the loading of D_2, which is now off and acting as a capacitance of about 20 pF. The second reason is the dispersion in the co-axial cable. From the simplest point of view, dispersion means an attenuation that increases with frequency. The negative going edge of the output pulse has to travel along an extra 4 m of cable on its journey to the oscilloscope, and suffer the consequent extra attenuation of its high frequency components. RG58C/U cable was used to obtain the results shown in Fig. 2.10, and this is the usual 50 Ω cable used for laboratory work which has an outside diameter of 5 mm. If a 2 m length of RG1788/U, a miniature 50 Ω cable of 1.8 mm outside diameter, is used instead, the negative going edge will be slower than that shown in Fig. 2.10 because of the greater dispersion in this smaller cable.

2.10 Conclusions

This chapter has discussed the problems of pulse generator circuit design in the particular case of the sampling pulse generator. This has to generate a very short pulse, of fixed amplitude and duration. The step recovery diode is certainly an excellent device for circuits of this kind, and it has been possible to propose an experimental circuit which uses a diode having step recovery properties, but working at a slower speed so that simple constructional and measurement techniques may be used.

Although the experimental circuit of Fig. 2.9 works at a speed which is, perhaps, two orders of magnitude slower than a real sampling pulse generator, it can still illustrate some of the problems of layout, and unexpected problems coming from component choice and test equipment, which turn up at the higher speed. Every version of a circuit of this kind will behave slightly differently, with different output pulse shape and different spurious signals appearing on the output. The length and amplitude of the pulse input, the grounding of circuit and test equipment, the earth-loops formed by the interconnection of test equipment: all these factors will influence what the experimentalist observes, and the whole point of the experimental work is to understand the real cause of every unexpected observation.

A real sampling pulse generator may, in some ways, avoid many of the difficulties that will be observed with the experimental circuit described here. This is because its final output pulse cannot be observed directly. It will be assumed that it has the correct amplitude and duration from the behaviour of the sampling gate, the subject of the next chapter, that the pulse generator controls. Another important point is that the sampling pulse generator and the sampling gate may, in a real sampling or digitising oscilloscope, be built together as one hybrid circuit: there will then be no problems introduced by a cable which has to carry the sampling pulse from generator to gate. In fact, as the next chapter will show, a sampling gate may well call for a controlling pulse that comes from a balanced source, in which case a balanced transmission line would be used to form the pulse, instead of the co-axial line shown in Fig. 2.9. Such a circuit is shown in the paper by Rush and Oldfield [12], and calls for a balanced reverse current source, in contrast to the single ended source shown in Fig. 2.9 as the single collector of Q_1. As a *circuit shape*, however, the balanced circuit is only the sum of two circuits, both having the shape of Fig. 2.9 but with one having all polarities of supplies and devices, with the exception of the centre of the circuit, D_2, reversed. This point is illustrated in Fig. 2.11 where a length, L, of balanced line, characteristic impedance Z_o, is shown connected to a balanced source of constant forward current for D_2, the step recovery diode, and a balanced pulsed source of reverse current, formed by Q_1 and Q_1'.

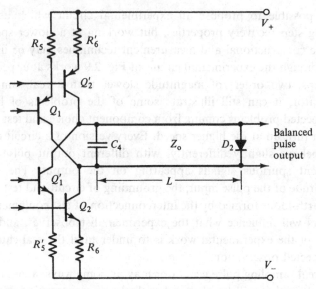

Fig. 2.11. *A balanced version of the sampling pulse generator. This is derived from Fig. 2.9. Components which correspond in both circuits have the same label in both, while those from the circuit of opposite polarity are marked with a dash. Bias networks and pulse inputs associated with the bases of the transistors have been omitted for simplicity.*

Notes

1 Jones, M. H., *A Practical Introduction to Electronic Circuits*, Cambridge University Press, Cambridge, second edition, 1985, p. 191.

2 White, E. L. C., British Pat. No. 595509, 1947.

3 Blumlein, A. D., British Pat. No. 528310, 1940.

4 Moll, J. L., Krakauer, S., and Shen, R., *Proc. IRE*, **50**, 43–53, 1962.

5 Hunter, L. P., *Introduction to Semiconductor Phenomena and Devices*, Addison Wesley, Reading, Mass., 1966, pp. 84–5.

6 Streetman, B. G., *Solid State Electronic Devices*, Prentice Hall, Englewood Cliffs, NJ, second edition, 1980, pp. 166–73.

7 Sze, S. M., *Physics of Semiconductor Devices*, John Wiley, New York, 1981, p. 108.

8 Kocsis, M., *High-speed Silicon Planar-epitaxial Switching Diodes*, Adam Hilger, London, 1976, pp. 50–5.

9 Pfeiffer, W., Zhang Zhizhong, and Zhou Xuan, *Arch. Elektr. Über.*, **36**, 39–42, 1982.

10 Becker, W., and Fomm, H., *Radio Ferns. Elektron.*, **25**, 812–3, 1976.

11 Note [1] above, pp. 138–9.

12 Rush, K., and Oldfield, D. J., *Hewlett–Packard J.*, **37**, No. 4, 9, Fig. 11, April 1986.

3
Sample and hold circuits

3.1 Introduction

In its simplest form [1], a sample and hold circuit has the circuit shape shown in Fig. 3.1. The signal input arrives through a 50 Ω co-axial line, to find a 50 Ω termination, and is then sampled periodically by means of the sampling gate. This gate is shown as a simple switch in Fig. 3.1.

The switch closes for a very short time, τ, short enough for it to be assumed that any change in the input signal over that time may be neglected. During this short time, the sampling capacitor, C_s, will begin to charge up towards the value that the input signal has at the instant of sampling. The time constant, T_{in}, associated with this charging process will be $25C_s$, because C_s will see a source impedance of 25 Ω when the switch is closed: the 50 Ω resistor in parallel with the 50 Ω input line.

When the switch is open, C_s is left providing an input to the voltage follower, A in Fig. 3.1. The output from the circuit is thus 'held' in between samples, provided the voltage follower can be made to have a high enough input impedance.

How the circuit actually behaves is mainly determined by the relative values of the input time constant, T_{in}, and the sampling time, τ. Three cases will be considered in the following three sections: $T_{in} \gg \tau$, $T_{in} \ll \tau$, and the particularly interesting case where C_s is replaced by an open circuited length of transmission line.

3.2 Performance when $T_{in} \gg \tau$

The sampling oscilloscope, the instrument discussed briefly in chapter 1, provides an example of a sample and hold circuit which operates with $T_{in} \gg \tau$. The signal input, in this case, is periodic and the time at which

27

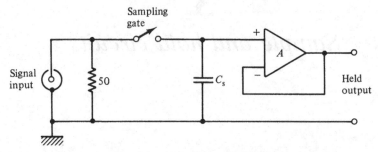

Fig. 3.1. *A sample and hold circuit.*

the sample is actually taken, t', is advanced, very slowly, step by step, so that the held output of the circuit shown in Fig. 3.1 is the integrated average of the input signal, $f(t)$, taken over time τ, that is

$$V_{\text{out}} = (1/\tau) \int_{t'}^{t'+\tau} f(t) \, dt. \tag{3.1}$$

An estimate of the effective bandwidth of the sample and hold circuit can be made by setting $f(t) = A \sin(\omega t)$ in equation (3.1). Completing the integration gives

$$V_{\text{out}} = -(A/\omega\tau)\{\cos[\omega(t'+\tau)] - \cos(\omega t')\} \tag{3.2}$$

which may be simplified by using the well-known identity,

$$\cos(\alpha) - \cos(\beta) = 2 \sin[(\alpha+\beta)/2] \sin[(\beta-\alpha)/2] \tag{3.3}$$

to give,

$$V_{\text{out}} = (2A/\omega\tau) \sin[\omega(t'+\tau/2)] \sin(\omega\tau/2). \tag{3.4}$$

Equation (3.4) shows that V_{out} is, as expected, a sine wave with the same frequency, ω, as the input sine wave, but transformed to the much slower time scale, t'. What is of primary interest, however, is the amplitude of this V_{out}. This amplitude is not A but, from equation (3.4),

$$|V_{\text{out}}| = (A/\pi f\tau) \sin(\pi f\tau) \tag{3.5}$$

where $f = \omega/2\pi$ has been substituted for convenience.

Equation (3.5) shows that the sample and hold circuit, shown in Fig. 3.1, has the frequency response shown in Fig. 3.2, when it is operating with $T_{\text{in}} \gg \tau$, and t' is changing very slowly. There will be an opportunity, later in this chapter, to check this experimentally. Fig. 3.2 has been drawn for the case where $\tau = 1$ ns. As would be expected, there is no response at all to a sinusoidal input which has a period equal to the sampling pulse width: the integral given as equation (3.1) is then zero. Similarly, there is no output for sinusoidal inputs having period τ/n, where n is any integer.

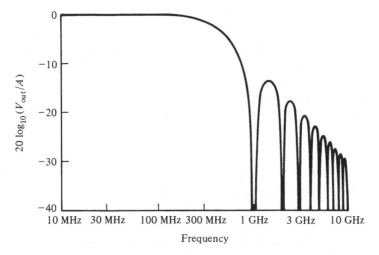

Fig. 3.2. *The response of the sample and hold circuit to sinusoidal inputs when* $T_{in} \gg \tau$, *the sampling time is advancing very slowly, and the sampling width,* τ, *is 1 ns.*

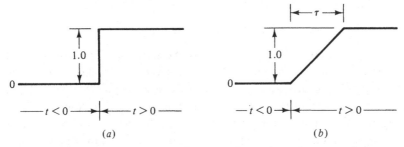

Fig. 3.3. *A Heaviside unit step function,* (a), *produces the output shown as* (b) *on the display of the sampling oscilloscope.*

As $[\sin(x)]/x = 0.707$ when $x = 1.39$, it is often stated [2] that the 3 db bandwidth of a sampling oscilloscope is, from equation (3.5)

$$B = 1.39/\pi\tau = 0.442/\tau \tag{3.6}$$

which would mean that a sampling pulse width of 1 ns would give a bandwidth of 442 MHz, as shown in Fig. 3.2. This can be a misleading point of view, however, because the frequency response shown in Fig. 3.2 does not belong to the simple world of linear network analysis, where everything has a slope of zero, 6 db/octave, 12 db/octave, and so on.

Just how different the sampling oscilloscope is, can be seen by considering equation (3.1) for another simple input function: the unit step function shown in Fig. 3.3(a). In this case the integration of equation (3.1) is so simple that it may be seen at once that the sampling oscilloscope display will be as shown in Fig. 3.3(*b*), where *attributed* time, *t*, has been used.

The result is that a perfect step input to the sampling oscilloscope is displayed as a linear rise, from zero to unity, duration τ. It might then be argued that the rise time, τ_r, of the sampling oscilloscope, taken from the 10% to the 90% points, is 0.8τ. Now using the often quoted [3] idea that bandwidth, *B*, and rise time, τ_r, are related by,

$$B = 0.35/\tau_r \qquad\qquad (3.7)$$

for a well-designed direct coupled amplifier, it would seem to follow that the bandwidth of a sampling oscilloscope is given by,

$$B = 0.35/0.8\tau = 0.468/\tau \qquad\qquad (3.8)$$

which is not all that different from equation (3.6). However, no linear amplifier can have a step response of the kind shown in Fig. 3.3(*b*), anymore than it can have a frequency response of the kind shown in Fig. 3.2. These are results which are specific to sampled data systems.

3.3 Performance when $T_{in} \ll \tau$

When the time constant, T_{in}, formed by the sampling capacitor, C_s, and the signal source impedance, in Fig. 3.1, is very much shorter than the sampling pulse width, τ, C_s has time to charge up to the signal level, always provided that the signal input frequency is well below $1/2\pi T_{in}$.

Operating with $T_{in} \ll \tau$ is the mode used in the sample and hold circuit at the input of a digitising oscilloscope. The input signal may now change dramatically from sample to sample, and the sample and hold circuit will still follow. In fact, the circuit is really operating in the track and hold mode, which, as Naegeli and Grau argue [4], has better noise and distortion performance than the sample and hold mode.

The frequency response of the circuit shown in Fig. 3.1, when $T_{in} \ll \tau$, is no longer that shown in Fig. 3.2, but a simple flat response out to a -3 db point at $f = 1/2\pi T_{in}$, followed by a -6 db/octave slope. Again, the experimental circuit, described towards the end of this chapter, will give an opportunity to demonstrate this.

In the digitising oscilloscope, the sample and hold circuit is not followed by a deflection amplifier of modest bandwidth, as it is in the sampling oscilloscope, but by a very fast ADC which must digitise the sample in, what may be, the very short time between samples. These very fast ADC circuits will be discussed in chapter 4.

When a very wide bandwidth is called for in a digitising oscilloscope, it is necessary to make the sampling pulse width, τ, very small indeed. To then make $T_{in} \ll \tau$ may be quite impossible if a simple capacitor is to be used in the circuit, as shown in Fig. 3.1. A very interesting way of getting around this problem is described in the next section.

3.4 An open circuit transmission line for C_s

Rush and Oldfield [5] have described a sample and hold circuit in which a length of transmission line, with both ends open circuited, is used in place of the capacitor, C_s, shown in Fig. 3.1.

Fig. 3.4 shows the essential details of this circuit. When the sampling gate, the simple switch shown in Fig. 3.4, closes, the signal is presented with a 25 Ω load. This is a perfect match to the signal source impedance, which is made up from the 50 Ω termination, shown in Fig. 3.4, in parallel with the 50 Ω cable that would be connected to the signal input socket.

It follows that, for the very first sample, a voltage equal to one half of the signal input will be developed across the input end of the 25 Ω line, shown in Fig. 3.4. This voltage step will propagate along the 25 Ω line, be reflected without change of sign at the open circuit, and return to the input end, charging the line up to the full signal voltage as it does so. If the switch in Fig. 3.4 is then opened, at exactly the instant that this reflection reaches the input end of the 25 Ω line, the line will be left charged to exactly the level of the signal input.

Subsequent samples will increase, or decrease, the voltage level held in the 25 Ω line by just the amount that the signal input changes between samples. This, of course, is only true when the variation in the signal level during the sampling time, τ, may be ignored.

The above argument implies that the sampling pulse width must be made exactly equal to the time taken for a signal to travel a distance $2L$ in the 25 Ω line, that is $\tau = 2L/v_p$. This is the same relationship that applied for the pulse generator circuits, Figs. 2.2(b), 2.9 and 2.11, discussed in the previous chapter. The pulse generator and sampling gate circuits now involve very similar transmission lines, and may be built together as one hybrid circuit. This is clear from the interesting photographs in Toeppen's paper [6] where both transmission lines can be seen in the hybrid circuit, these being only a few millimetres in length because a sampling pulse width of only 100 ps is being used in this case.

What kind of frequency response does the sample and hold circuit shown in Fig. 3.4 have? The answer to this question may be seen by considering a sinusoidal signal input with a period exactly equal to τ, and thus also equal to $2L/v_p$. This means that the 25 Ω line is one half

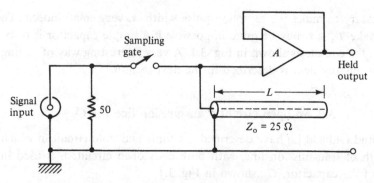

Fig. 3.4. *A sample and hold circuit using a length of open circuited line instead of a sampling capacitor.*

wavelength long. No matter at what points on the sinusoidal input signal the sampling gate closed and opened, the result would always be the same: the 25 Ω line behaves as a high Q parallel resonant circuit and there would be no constant voltage level component to act as an input signal for the low frequency voltage follower amplifier, A in Fig. 3.4. This means that the circuit of Fig. 3.4 has a frequency response of the same form as that shown in Fig. 3.2. The effective bandwidth will be of the order of that given by equation (3.6), that is $0.442/\tau$. However, there is a very important improvement in the performance in that the restriction on the change in input signal level from sample to sample, which was required in order to obtain equation (3.6), no longer applies for the new circuit, Fig. 3.4. No matter how great the change in signal input, the 25 Ω line shown in Fig. 3.4 will be left charged up to the correct level. The problem now is to digitise this analog information before it changes again, but that is the problem considered in the next chapter. This chapter continues with a look at the hardware realisation of sample and hold circuits.

3.5 The realisation of a sampling gate

So far, the sampling gate has been shown, in Figs. 3.1 and 3.4, as a simple switch. In practice, a very fast electronic switch is needed, and one which can be operated by the sampling pulse.

As an example, consider the idea of using a single n-channel enhancement MOST as a sampling gate. This idea is shown in Fig. 3.5. Three obvious problems can be seen at once.

1 The resistance of the device, when it is on, should be negligible compared to the signal source impedance, which in this case is 25 Ω. No fast MOST will satisfy this condition.

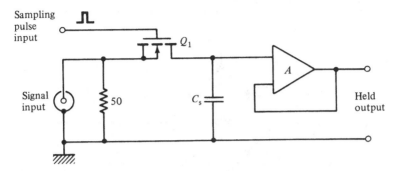

Fig. 3.5. *A single MOST does not make a good sampling gate.*

2 The impedance of the device must be very high indeed when it is off, particularly when C_s is small. An MOST will have a finite drain to source capacitance.

3 There will be considerable breakthrough of the sampling pulse into the signal input circuit, and across C_s, because of the finite gate to source and gate to drain capacitances.

These three problems are the main concern with all very high frequency sampling gate circuits. A review by Akers and Vilar [7] suggests that fast response diodes may be the most useful devices, at present, for the realisation of sampling gates, because these can have a very low on-resistance, very high off-impedance, and several matched diodes may be arranged in various configurations to provide good isolation between sampling pulse input and signal input. For this reason, the remainder of this chapter will concentrate upon these sampling gate circuits which use diodes, but it must always be borne in mind that this is only a fraction of a large and rapidly developing field in electronic circuit design. Among the many devices reviewed by Akers and Vilar [7], an obviously strong candidate for the future is the opto-electronically controlled switch. This idea comes from some early work by Auston [8], using the photo-conductivity of silicon. Indium phosphide opto-electronic switches have been used as sampling gates, at picosecond speeds, for the direct digital processing of radar signals [9].

3.6 Diode sampling gates

In the previous section, low on-resistance was listed as the first requirement for a sampling gate device. The solid state diode has good possibilities from this point of view, because it has a small signal resistance

$$\mathrm{d}V_{\mathrm{F}}/\mathrm{d}I_{\mathrm{F}} = (kT/e)/I_{\mathrm{F}} \qquad (3.9)$$

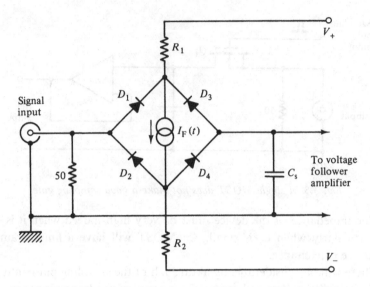

Fig. 3.6. *A balanced diode bridge sampling gate. This should show
negligible breakthrough of the sampling pulse, $I_F(t)$, into the signal
circuit.*

when it is forward biassed with a current I_F. As kT/e is 25 mV at room
temperature, this means that a forward resistance of a few ohms may be
obtained with a forward current of only a few milliamps.

The diode impedance can be very high when the diode is reverse
biassed, so the second point listed in the previous section is also well
satisfied.

It is the third point which influences the circuit shape of diode sampling
gates more than anything else: the isolation which is needed between the
sampling pulse input to the gate and the low level signal circuit.

Fig. 3.6 shows a first suggestion for a diode sampling gate. This uses
four diodes connected in a bridge arrangement. The diodes are normally
all reverse biassed by the connection to positive and negative supplies *via*
R_1 and R_2, which are both made high and equal in value. This is the state
when the sampling gate is off and the isolation between the signal input
and the sampling capacitor, C_s, is provided by the high impedance and
small capacitance of the reverse biassed diodes.

The gate is turned on by means of a sampling current pulse, $I_F(t)$.
Because of the symmetrical bridge connection of the four diodes, this
current is contained within the diode bridge, except for a small fraction
which must flow in the resistors R_1 and R_2, and there should be no fraction
of $I_F(t)$ flowing into C_s or into the signal input circuit. To obtain this
result, excellent matching must be achieved between the four diodes. The

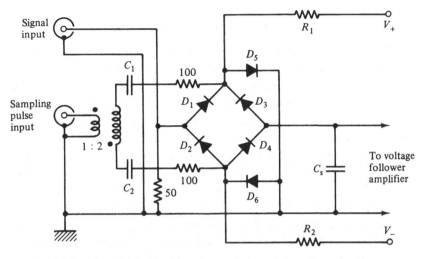

Fig. 3.7. A balanced series/shunt gate using six diodes. Some details are now shown for the sampling pulse input to this circuit.

stray capacitances associated with top and bottom nodes of the diode bridge, as it is shown in Fig. 3.6, must also be matched as closely as possible.

3.7 A series/shunt gate

A considerable improvement in the performance of the gate shown in Fig. 3.6 is obtained if two more diodes are added. These are shown in Fig. 3.7 as D_5 and D_6. The aim is to improve the isolation between the signal input and the sampling capacitor, C_s, when the gate is off. This is done by having D_5 and D_6 forward biassed when the gate is off, so that they present a very low impedance at the upper and lower nodes of the diode bridge, which previously had only high impedances, R_1 and R_2, connected to them. In Fig. 3.7, the signal input is now isolated from the sampling capacitor by the large attenuation that is effected through the very high impedances of D_1 and D_2, followed by the very low impedances of D_5 and D_6.

But the addition of D_5 and D_6 has the further advantage of clamping the upper and lower nodes of the diode bridge to approximately ± 0.7 V, in the case of silicon diodes, so that R_1 and R_2 can now be made really high values without involving large voltage excursions across the diodes themselves. The first task of the sampling pulse input is to turn D_5 and D_6 off, forcing the upper and lower nodes of the diode bridge to go from the ± 0.7 V state to the ∓ 0.7 V state, where D_1, D_2, D_3, D_4 are all on.

3.8 Connecting the sampling pulse generator

Fig. 3.7 also begins to show some details of the way in which the sampling pulse generator might be connected to the sampling gate. Ideally, a balanced, and truly floating, pulsed current source is needed, as shown in Fig. 3.6. This was discussed briefly in the last section of chapter 2, where Fig. 2.11 was presented as a circuit shape for such a balanced sampling pulse generator.

Experimentally, it is most convenient if a simple sampling pulse generator may be used, and this will have an unbalanced 50 Ω co-axial output. This is true of any simple laboratory pulse generator, and also true of the experimental circuit described in the previous chapter: Fig. 2.9.

To use such an unbalanced pulse source as a sampling pulse generator for the balanced diode gate, a pulse transformer has been used in Fig. 3.7, which transfers the pulse from the 50 Ω co-axial system into a balanced 200 Ω transmission line system. Use is then made of the very low forward resistance of the diodes, when compared to the higher impedance of the balanced transmission line system, to make a fair approximation to a current source for switching the diodes on and off.

Such a balanced to unbalanced transformer is usually termed a 'balun'. This is not a simple transformer but a very wide bandwidth, high frequency device, and its construction will be dealt with in the next section. In circuit representation, however, a balun is usually shown by the normal transformer symbol.

A transformation ratio of 1:2 is the most simple one to obtain, and this means an impedance ratio of 1:4, so that the secondary of the balun shown in Fig. 3.7 is terminated with 200 Ω. This is done using two separate 100 Ω resistors so that there is a balance in the unavoidable stray capacitance to ground.

It is very important to understand the reason for the capacitors C_1 and C_2 in Fig. 3.7. These are essential because there can be no d.c. level across the secondary of a transformer. The fact that some balun designs are isolating transformers, like the one symbolised in Fig. 3.7, and others involve a d.c. path between primary and secondary, has nothing at all to do with this point. C_1 and C_2 are needed because D_5 and D_6 clamp the upper and lower nodes of the diode bridge at ± 0.7 V for most of the time. Just how much of the time depends upon the mark to space ratio of the sampling pulse generator. C_1 and C_2 charge up to accommodate this change in mean d.c. level between the secondary of the balun and the upper and lower nodes of the diode bridge. Two capacitors are used for the same reason, discussed above, that two 100 Ω resistors are used instead of a single 200 Ω one. These capacitors would be made quite large, and, of course, equal, so that the time constant, $100 \,\Omega \times C_1$, would be greater than the longest time expected between samples.

3.9 A design for a balun

The first step towards the construction of an experimental diode sampling gate circuit is to construct the balun that will be needed between this circuit and the sampling pulse generator. This balun will be a pulse transformer, that is a device with a very wide bandwidth, and must be constructed from transmission line.

As an introduction to this method of pulse transformer construction it is useful to consider the simple inverting pulse transformer that consists of a length of co-axial transmission line wound upon a ferrite toroid. At one end of this line, the outer sheath is grounded and the inner conductor is connected to the pulse source, in the usual way. At the other end of the line, because of the large inductance of the winding, which is common to both inner and outer conductors, the connections may be *reversed*: the inner conductor may be grounded while the outer conductor used as the source of a pulse, identical to the input pulse but of opposite sign. This kind of transformer has been known for a very long time [10].

The simplest way to make a balun transmission line transformer is to take two of the inverting transformers, that have been described above, and connect their inputs in parallel on one side and in series on the other. For example, 100 Ω co-axial cable could be used to provide a 50 Ω to 200 Ω balun transformer of this kind. Using 100 Ω twisted pair, or balanced line, would be even better. This method of constructing a balun may have been first described by Talkin and Cuneo in 1957 [11] and a paper [12] has reviewed developments in technique, although it does not mention an important contribution by Hilberg [13], who deals with the problems that can arise when transformers are made to operate in both the transmission line mode and the conventional coil mode, simultaneously.

The balun transformer which is recommended for the experimental circuit of this chapter, is shown in Fig. 3.8. This uses four of the inverting transformers that were described above, using four equal lengths of 50 Ω twisted pair transmission line wound on four ferrite toroids. For simplicity, only a single turn, and a single twist, are shown on each toroid in Fig. 3.8. In practice, 500 mm of twisted pair line is used on each core, these being 25 mm diameter, high permeability, high frequency, ferrite [14]. The winding is widely spaced. A 50 Ω twisted pair line is easily made

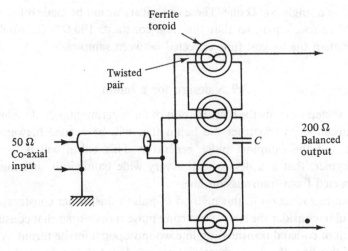

Fig. 3.8. *One possible way of realising the balun, shown in Fig.* 3.7, *using four ferrite toroids and four lengths of* 50 Ω *line made from twisted pairs. Capacitors* C_1 *and* C_2, *in Fig.* 3.7, *are now replaced by a single one,* C.

from two 0.4 mm polyurethane coated wires, twisted to a 5 mm pitch. As Fig. 3.8 shows, the four 50 Ω lines are connected in series/parallel on the 50 Ω side, to give a 50 Ω input impedance, while they are all four connected in series, on the output side, to give the 200 Ω balanced output. This configuration has a very great advantage over the simpler one, using just two 100 Ω lines, in that the balun shown in Fig. 3.8 is an isolating transformer: there is no d.c. path between input and output.

A more expensive way of making the balun shown in Fig. 3.8, that certainly gives better very high frequency performance, is to use four lengths of miniature 50 Ω co-axial cable, instead of the twisted pairs, and thread ferrite beads on each length [15]. This gives a linear layout to the balun and reduces the shunt capacitance that must be associated with the toroidal windings of the previous version.

3.10 An experimental sample and hold circuit

The experimental circuit for this chapter is shown in Fig. 3.9. This uses the six diode configuration, previously shown in Fig. 3.7. The CA3019 device is used for this: an array of six diodes, in monolithic silicon, in which four diodes are internally connected as a bridge, and the other two are left free. The diodes are numbered in Fig. 3.9 so that they correspond to the data sheet numbering [16]. The sampling pulse input comes directly from the

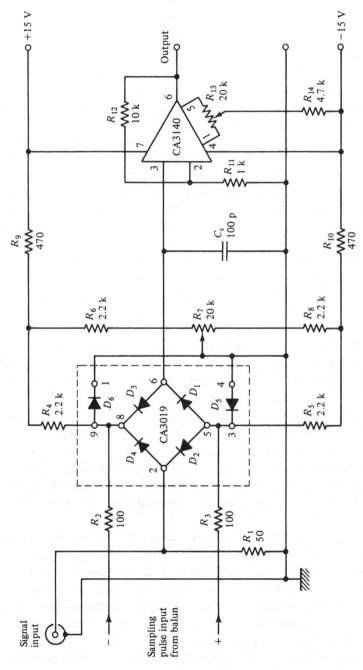

Fig. 3.9. The experimental sample and hold circuit. Note that pin 7 of the CA3019 must be connected to −7.5 V and, as with pins 4 and 7 on the CA3140, decoupled to ground.

200 Ω balanced output of the balun, shown in Fig. 3.8. The capacitor C, shown in Fig. 3.8, should be a ceramic 0.33 μF, 50 V, or a larger value if this is available in a small size with low inductance.

The amplifier which is needed to produce a useful output from the voltage that is held on the sampling capacitor, C_s, is provided by the CA3140 in Fig. 3.9. This is connected to have a gain of 11 by means of the feedback network R_{12} into R_{11}. The CA3140 is a very well-established operational amplifier [17] that has MOS input transistors, and thus draws negligible current from the small capacitor, C_s. Attention must be given to the layout of the circuit board around the top end of C_s, and all the connections to this, to ensure that there is negligible leakage current from any source into this sensitive area. This point can be easily checked, when the circuit is operating, by removing the sampling pulse input to the primary of the balun and watching how rapidly the d.c. level, at the output of the CA3140, drifts away from the zero level previously set by the nulling operation. Even when C_s is as small as 100 pF, this drift, which can be positive or negative, should be only of the order of ± 1 V/s.

3.11 Experimental work

Because the MOS input devices in the CA3140 cause this to have a fairly large input offset voltage, it is necessary to null the output of the circuit shown in Fig. 3.9, with C_s temporarily short circuited, before starting work. This is done by adjusting R_{13}.

A sampling pulse of $+4$ V amplitude, at least 10 kHz repetition rate, and between 20 ns and 100 ns duration, should then be supplied to the input side of the balun, shown in Fig. 3.8. This pulse may be provided by the experimental circuit of the previous chapter, Fig. 2.9, or from a laboratory pulse generator. The short circuit should be removed from C_s and a 50 Ω termination connected to the signal input socket, shown in Fig. 3.9.

First, view the d.c. level at the output with an oscilloscope. This will no longer be zero, as it was after the null was effected with R_{13}, but offset by some small amount which will be found to depend upon the sampling pulse width. The reason for this is the unbalanced dynamic behaviour of the diodes. Even if all six diodes were perfectly matched from a d.c. point of view, the stray capacitance, intrinsic to their monolithic structure, is unbalanced, and the switching characteristics must differ. This means that a small amount of charge will be transferred into or out of C_s during the rise time of the sampling pulse. The opposite occurs during the fall time. It is impossible for these two dynamic events to cancel one another out, and a small positive or negative charge will be transferred to C_s even when the signal input is zero. This can be corrected by a very small adjustment of the forward bias on D_6 and D_5, and thus the reverse bias on the other four diodes. This adjustment is made by means of R_7, for each particular value of sampling pulse width.

Why does the sampling pulse width influence the output offset corresponding to zero input signal? The answer to this may be found by observing how the various ground connections and lengths of cable used in the experiment influence this offset. The rising edge of the sampling pulse, and the falling edge, set up disturbances that persist for some microseconds as they are reflected and transmitted around the various transmission paths within the system. Changing the relative time position of these two events, rise and fall, by changing the pulse width, changes the way in which these spurious signals interfere with one another, and this, in turn, must have a slight effect upon the way in which the diode gate turns off at the end of the sampling pulse. These problems do not, of course, arise in the final form of a manufactured digitising or sampling oscilloscope, because the sampling pulse width is fixed, and so are the relative positions of the various parts of the system.

3.12 Measuring the frequency response

Once the experimental circuit has been nulled correctly under zero signal input conditions, the input termination may be replaced with a simple radio frequency signal generator, and the behaviour of the sample and hold circuit checked for a sine wave input.

At present, the value of C_s is 100 pF, so that T_{in} is 2.5 ns: very much smaller than the sampling pulse width, which will be around 50 ns. This means that the circuit is operating in the $T_{in} \ll \tau$ regime, discussed in section 3.3, and should have a frequency response out to $1/(2\pi \times 2.5 \times 10^{-9})$ Hz, or 63.6 MHz. This can be checked using very simple equipment, because the actual output from the CA3140, in Fig. 3.9, is of very limited bandwidth: only about 300 kHz. If a simple analog oscilloscope is connected to the output of the CA3140, and its timebase is triggered at the repetition rate of the sampling pulse generator, which is around 10 kHz, the envelope of the sampled data will be observed on the analog oscilloscope, and this will, in fact, look like a sine wave whenever the input frequency is close to a harmonic of 10 kHz. Naturally, it is not easy to hold the frequency of a simple signal generator close to, for example, exactly 300×10 kHz, or 30 MHz. The repetition rate of the pulse generator will also be subject to drift. In practice a sinusoidal looking output will be observed only every now and again, as the input frequency is varied, but the *envelope*, or *amplitude*, of this sampled output may easily be measured so that the bandwidth of 63.6 MHz can be checked. The value of C_s may then be changed to 200 pF, or 50 pF, and the check repeated.

The other important point to check is the overall gain of this sample and hold system, and its noise level. An RF input of 10 mV peak to peak should give an output with an envelope of 110 mV peak to peak, because there should be negligible loss across the diode gate followed by a gain of 11 in the CA3140. It is this overall gain of 11 that should drop by 3 db at 63.6 MHz.

The noise level at the output of the CA3140 depends very much upon the laboratory environment: this is a very wide bandwidth system of high sensitivity, and is laid out in a rather open way. However, on a timebase of, say, 1 ms/div., the noise should appear to be only a few millivolts. The breakthrough of the sampling pulse is discussed in section 3.13.

It is now interesting to examine the other mode of behaviour that was discussed in section 3.2: the case when $T_{in} \gg \tau$. To do this the value of C_s is simply increased several orders of magnitude. For example, making $C_s = 10$ nF will increase T_{in} to 250 ns. Working with a sampling pulse of 50 ns, a frequency response like the one shown in Fig. 3.2 should be obtained with the experimental circuit, but the first null in the response should be at 20 MHz and, from equation (3.6), the 3 db bandwidth should be at 8.84 MHz.

When this is checked, using the same technique that was described at the beginning of this section, something, perhaps, unexpected will be observed. Although the frequency response will be found to follow the interesting form shown in Fig. 3.2, and the complicated periodic responses and nulls out to the higher frequencies will be clearly observed, the entire response will appear to show a rapidly fluctuating overall gain as the input frequency is increased slowly over the expected pass bands.

This behaviour is a demonstration of the constraint that was put upon the rate at which t', the sampling time, could change when equation (3.1) was used to derive equation (3.4). The behaviour shown in Fig. 3.2 will only be found in the experimental circuit when the signal input frequency is very close indeed to a harmonic of the sampling pulse generator repetition rate. The experimental circuit is then behaving in exactly the same mode as a classical sampling oscilloscope: the sampled output is equivalent to a stroboscopic view of the high frequency input.

Finally, it is interesting to attempt to model the idea of using an open circuited length of transmission line for C_s. This idea was discussed above in section 3.4. The 25 Ω line which is needed is easily made by using two 50 Ω co-axial cables in parallel, with both inner conductors connected to pin 3 of the CA3140, in Fig. 3.9, and both outer conductors taken to ground. C_s is, of course, removed entirely. For a 50 ns sampling pulse, the two lines will have to be about 5 m in length. The same frequency response seen in the previous test, with $C_s = 10$ nF, should then be observed but, this time, the restriction on the rate of change of t', and the rapid fluctuations in overall gain, observed previously as the input frequency was changed, should disappear.

3.13 Dynamic range

Up to now, all the tests on the experimental sample and hold circuit have been made with an RF input of only a few millivolts. Looking at Fig. 3.9, it is clear that there must be some upper limit to the signal input level because once this begins to approach the forward drop across any of the diodes the gate can no longer be operating in a simple way.

It is useful to combine a test of the d.c. input level handling capabilities of the circuit with an observation of the breakthrough of the sampling pulse into the input circuit and the output circuit. To do this, a simple analog oscilloscope should be used to view the d.c. level at both the input and the output of the circuit shown in Fig. 3.9, and the timebase of this oscilloscope triggered from the trigger output of the pulse generator being used to provide the sampling pulse. If the timebase speed is then set to, say, 1 μs/div, the breakthrough of the sampling pulse into the input circuit, and at the output of the CA3140, will be seen as spikes. These spikes should be only about 100 mV on either channel, and are, of course, a distortion of the sampling pulse by the limited bandwidth of the oscilloscope, and the CA3140. Nevertheless, a relative measure of the sampling pulse breakthrough, and how it changes, can be made.

A d.c. level may now be established at the input by connecting the input termination to a variable voltage power supply, say ±30 V *via* a 500 Ω resistor. This should make an input voltage of up to ±1.5 V possible. As the d.c. input is increased, the output should also increase, 11 times faster because of the gain of the CA3140, and the sampling pulse breakthrough will change because the balance of the diode bridge is being spoilt as the input voltage level departs from zero. Apart from this, the gate will appear to continue working until the input level reaches two forward diode voltages in magnitude, that is 1.2 V or more. Real confirmation that the gate is still turning off properly may be obtained by disconnecting the sampling pulse and checking that the output level of the CA3140 still drifts slowly, one way or the other, despite the high d.c. level at the gate input. Note that this result means that a larger dynamic range may be obtained by using gallium arsenide diodes [18], while Schottky barrier diodes give less dynamic range than silicon.

3.14 Conclusions

After a general discussion of sampling gates and the kind of overall frequency response that can be expected from sampling systems, this chapter took the six diode series/shunt gate as an experimental project for circuit design work.

This involved a short digression, in section 3.9, to consider the construction of the balun that was needed to connect the sampling pulse generator to the diode gate. In a real digitising oscilloscope, this balun would not be needed because the combination of sampling pulse generator and sampling gate becomes one single hybrid circuit. However, it is much easier to observe what is really happening, and to vary the circuit parameters experimentally, when the system is split up into convenient parts and connected up with co-axial cable. The earth-loop problems and matching problems that then come up all add to the interest and value of the experimental work.

Another radical difference between the experimental circuit described here, Fig. 3.9, and a real digitising oscilloscope, is the use of a much longer sampling pulse. This is essential if experimental work is to be done using fairly simple methods of construction and test equipment. The system described here is a scaled-up version of the real thing: lengths of transmission line are several metres in length, instead of several millimetres. The principles involved are just the same, however.

The message coming from the experimental work of this chapter should be about *time*. The experimental circuit accepts signals with frequencies approaching 100 MHz at its input, and then reproduces these signals, accurately, on a much slower time scale. Once the sampling gate has been passed, the hardware involved is all low frequency: a simple operational amplifier and a general purpose laboratory oscilloscope.

The same message is found inside the case of a digitising oscilloscope. The input circuits belong to the world of microwave electronics and occupy, perhaps, 5% of the volume. There is then a very fast ADC, the subject of the next chapter. After that, all the electronics is rather slow and very cheap: digital data processing, memory and display.

Notes

1 Jones, M. H., *A Practical Introduction to Electronic Circuits*, Cambridge University Press, Cambridge, second edition, 1985, pp. 258–9.

2 Carlson, R., Krakauer, S., Magleby, K., Monnier, R., Van Duzer, V., and Woodbury, R., *IRE WESCON Conv. Rec.*, 3, Part 8, 44–51, 1959.

3 Valley, G. E., and Wallman, H., *Vacuum Tube Amplifiers*, Dover, New York, 1965, p. 80.

4 Naegeli, A. H., and Grau, J., *Hewlett–Packard J.*, 38, No. 11, 6–17, Dec. 1987.

5 Rush, K., and Oldfield, D. J., *Hewlett–Packard J.*, 37, No. 4, 4–11, April 1986.

6 Toeppen, D. E., *Hewlett–Packard J.*, 37, No. 4, 33–6, April 1986.

7 Akers, N. P., and Vilar, E., *Proc. IEE*, 133A, 45–9, 1986.

8 Auston, D. H., *Appl. Phys. Lett.*, 26, 101–3, 1975.

9 Yao, I., Diaduik, V., Hauser, E. M., and Bouman, C. A., 'High-speed opto-electronic track and hold circuits in hybrid signal processors for wide band radar'. In: *Picosecond Electronics and Opto-electronics*, Eds. Mourou, G. A., Bloom, D. M., and Lee, C.-H., Springer, Berlin, 1985, pp. 207–11. (Springer Series in Electrophysics: Vol. 21.)

10 Winningstad, C. N., *IRE Trans. Nucl. Sci.*, NS-6, 26–31, 1959.

11 Talkin, A. I., and Cuneo, J. V., *Rev. Sci. Instr.*, 28, 808–15, 1957.

12 Yamazaki, H., Homma, A., and Yamaki, S., *Rev. Sci. Instr.*, 55, 796–800, 1984.

13 Hilberg, W., *IEEE Trans. Mag.*, MAG-6, 667–8, 1970.

14 Mullard FX3312 Ferroxcube toroids are a good choice. So are the Siemens R25 toroids in type N30 Siferrit.

15 O'Dell, T. H., *Electronics and Wireless World*, 94, No. 1630, 767, Aug. 1988.

16 RCA Data Bulletin File No. 236. The LM3019 from National Semiconductor is an equivalent device.

17 RCA Data Bulletin File No. 957.

18 Gibson, S. R., *Hewlett–Packard J.*, 37, No. 2, 4–10, Feb. 1986.

4

Comparator circuits

4.1 The analog to digital converter (ADC)

The sub-system which follows the sample and hold circuit, and its sampling pulse generator, in a digitising oscilloscope will be an ADC. In this particular case, the function of the ADC is to convert the output from the sample and hold circuit, which is an analog signal carried by a single channel, into a digital representation of this analog signal, carried in parallel by, typically, eight lines. An eight bit representation would be only just adequate for a digitising oscilloscope because this would give 256 discrete levels, meaning that a vertical display, 8 cm high, would have a vertical resolution of about 0.3 mm. This is about the same as a typical CRO spot diameter. More bits would be needed to handle the display offset, the sign of the signal, and the sensitivity range, but only the circuit problems of the ADC are considered in this chapter.

ADCs are not only found in digitising oscilloscopes. Any instrumentation problem that calls for an interface between a transducer or a sensor, nearly all of which give out an analog signal, and a digital computer, will call for an ADC. There is also another field of application, as Gordon [1] has pointed out in an important review. This is the field of communications where digital techniques are taking over more and more, in telephony and in television.

In its most basic form, an ADC may be considered as a sub-system, or device, of the kind shown in Fig. 4.1, where it is assumed that the analog input is always positive, lying between zero and V_{ref}, and is to be converted into an eight bit binary coded output. This device would usually be a single silicon integrated circuit.

There are a very large number of ways in which the ADC function may be realised. Gordon [1] identified twelve distinct types, six of which may

46

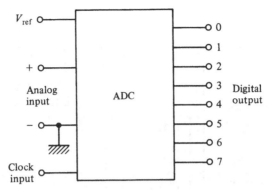

Fig. 4.1. *The analog to digital converter* (*ADC*).

be classified as slow and six fast. The slow designs use counting techniques to generate the digital representation of the analog signal. They all rely on converting the analog voltage into a time: for example, the time for a current to charge a capacitor. Fast designs use some kind of parallel data processing technique.

Nearly all ADC designs involve *comparators*. These circuits are the subject of this chapter and, as the name implies, their function is to compare a time varying signal with either another signal, or with some reference level. The comparator determines the instant when a charging capacitor has a certain voltage across it, or whether a voltage is above or below a certain level.

4.2 The flash ADC

Of all ADC designs, the so-called flash ADC is the fastest and is the one used in all fast digitising oscilloscopes. It is also the design which makes the most use of comparator circuits.

Fig. 4.2 shows the kind of structure that a flash ADC would have. The origin of this idea, as Kandiah has pointed out [2], is very old indeed. It is found in the 1940s, in the earliest pulse height analysers of nuclear physics instrumentation.

Fig. 4.2 has been drawn so that it corresponds to Fig. 4.1: again the eight bit digital output has been assumed. The flash ADC works by having 255 comparators to detect exactly where the analog input signal level lies, relative to the 256 levels that an eight bit digital output is able to represent. These comparators are represented by simple operational amplifier symbols in Fig. 4.2. An operational amplifier can, of course, be used as a comparator, but certainly not as a fast one.

The outputs from the 255 comparators, shown in Fig. 4.2, represent the

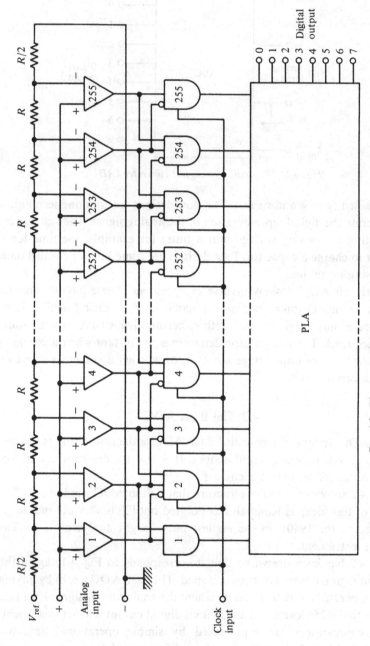

Fig. 4.2. The internal details of a typical flash ADC.

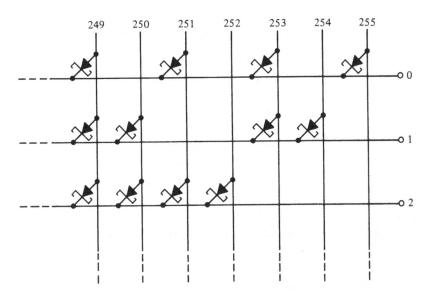

Fig. 4.3. *The PLA in Fig.* 4.2 *may be realised by means of a simple Schottky diode matrix. Here the details of the top right-hand corner of such a matrix are shown.*

analog signal level in what is called a 'thermometer code' [3]. This term is used because, as a glance at Fig. 4.2 will show, all the comparators which take a reference voltage on their inverting inputs that lies below the analog input signal level, will have their outputs high. In contrast, all the comparators which have high reference levels will have their outputs low.

The thermometer code, described above, has to be converted into the standard binary number representation that is needed at the output of the ADC. There are many ways of doing this encoding, and the method presented here has been adapted from the paper by Peterson [4], which described one of the earliest realisations of a flash ADC as a single monolithic silicon integrated circuit. This uses the 255 AND gates, shown in Fig. 4.2, to convert the thermometer code from the 255 comparator outputs into a '1 out of 255' representation. The only AND gate that can have a '1' at its output is the one which has the input, reading from left to right in Fig. 4.2, '011'. That is exactly the 'thermometer' reading. The '1 out of 255' code is then easily converted into the eight bit binary code. This could, for example, be done by means of a programmable logic array (PLA), and the simplest PLA would be a diode matrix. For completeness, a small part of such a diode matrix is shown in Fig. 4.3.

4.3 How system considerations influence circuit design

Having gone into some detail about the realisation of one particular kind of ADC, and also one particular kind of encoding structure for its output, it is now possible to bring up a point which is of crucial importance to electronic circuit design in general. This is the way in which system considerations influence circuit designers and play a major role in the kind of circuit shape at which they may finally arrive.

In the particular case being considered here, this discussion can begin by bringing in a very simple feature of the ADC that is shown in both Figs. 4.1 and 4.2. This is that the ADC must have a 'clock input'. The function of this clock input signal is to enable the ADC during the time that its analog input is held constant. In the digitising oscilloscope this time is after the sampling pulse, and after the sample and hold circuit has settled down to its new output value. The clock signal would be a square wave, of alternately '0', when the sampling pulse would be generated and the new held sampled value would be established, and then '1', when all the AND gates, shown in Fig. 4.2, would be enabled and the eight bit digital output would be available at the output of the ADC.

Now the above is only one way of realising this essential demand, of the complete digital system, that the output of the ADC is only valid when the analog input is constant. Having decided to solve the problem in this way, the circuit designer's attention would be directed towards the circuit detail, or circuit shape, of the AND gates in Fig. 4.2, and deciding how the enable function could best be implemented.

It is obvious, however, that the enable function could also be done by designing either the comparator circuits or the PLA, in Fig. 4.2, so that these circuits could accept the clock input signal. In fact, one of the most interesting published designs for a flash ADC, and one of the very first eight bit video ADCs to be made as a single integrated circuit [4], has the clock signal fed to comparators, AND gates, and the output encoding circuits. In this way it proved possible to make the ADC behave as its own sample and hold. It is clear that this overall system influence on the details of circuit design is a point of great importance, and it is fortunate that the flash ADC problem can illustrate it so well. To give this illustration, the remainder of this chapter considers simple comparator circuits briefly and then the evolution of the kind of comparator circuit design that has taken system considerations into account. This leads to the theory of such circuits and work with an experimental circuit.

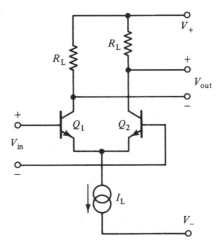

Fig. 4.4. *The simple long tailed pair circuit. A first step towards a good circuit shape for a comparator.*

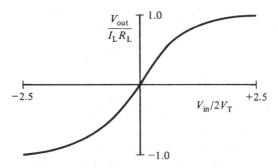

Fig. 4.5. *The transfer function for the circuit shown in Fig. 4.4.*

4.4 The simple comparator

A simple comparator could be made by using a device like an operational amplifier, and then omitting any feedback. The input circuit of such a simple comparator would almost certainly be the well-known, long tailed pair circuit, shown in Fig. 4.4, and this can be taken as a starting point in the evolution of a really fast kind of comparator circuit.

Fig. 4.4 has the transfer function shown in Fig. 4.5. This follows [5] from the equation for the collector current of a bipolar transistor. In general:

$$I_{\text{C}} = I_{\text{CBO}} \exp (V_{\text{BE}}/V_{\text{T}}) \tag{4.1}$$

where,

$$V_{\text{T}} = kT/e \tag{4.2}$$

which is about 25 mV at room temperature.

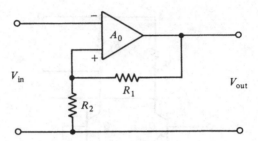

Fig. 4.6. *Positive feedback applied across a simple operational amplifier can produce a transfer function of the double valued form shown in Fig. 4.7.*

Because,

$$V_{in} = V_{BE1} - V_{BE2} \tag{4.3}$$

$$V_{out} = (I_{C1} - I_{C2}) R_L \tag{4.4}$$

and

$$I_L = I_{C1} + I_{C2} \tag{4.5}$$

the identity,

$$\tanh(x) = [\exp(x) - \exp(-x)]/[\exp(x) + \exp(-x)] \tag{4.6}$$

may be used to obtain,

$$V_{out}/I_L R_L = \tanh(V_{in}/2V_T) \tag{4.7}$$

which is the function shown in Fig. 4.5.

Obviously, the simple circuit shown in Fig. 4.4 would need additional output stages to give it more gain and both to level shift the output and make it single ended. These additions would not radically change the form of the transfer function, shown in Fig. 4.5, and it is with reference to this function that some obvious problems can be identified.

4.5 The use of positive feedback

The most striking feature of Fig. 4.5 that needs some attention, if the circuit of Fig. 4.4 is to lead to a useful comparator circuit, is the absence of any clear threshold input voltage at which the output voltage changes rapidly from the low state to the high. An increase in the circuit gain would tend to produce a sharper threshold, and it is from this point of view that the idea of applying positive feedback to increase the gain seems a sensible direction for the circuit designer to choose [6].

For example, the circuit idea shown in Fig. 4.6 uses positive feedback across a standard operational amplifier. As the feedback fraction is

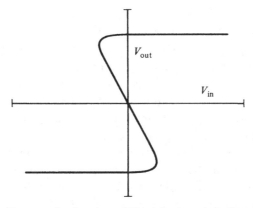

Fig. 4.7. *The transfer function of Fig.* 4.6 *when* $A_0 R_2/(R_1 + R_2) > 1$.

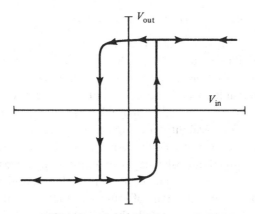

Fig. 4.8. *The double valued transfer function of Fig.* 4.7 *cannot be followed. The circuit of Fig.* 4.6 *exhibits a trigger action and hysteresis.*

increased, the transfer function changes from the form shown in Fig. 4.5 to give a higher and higher gain, passing though $+\infty$ to change, discontinuously, to $-\infty$ at the critical point where the overall loop gain, $A_0 R_2/(R_1 + R_2)$, is unity. If the amount of positive feedback is increased still further, the transfer function takes on the interesting form shown in Fig. 4.7, a feature of electronic circuits with positive feedback that was first clearly explained by Williams [7].

A circuit cannot, of course, operate smoothly along a multivalued transfer function of the kind shown in Fig. 4.7. The circuit now has a trigger action and exhibits hysteresis, as shown in Fig. 4.8. In fact, Fig. 4.6 is often referred to as a 'Schmitt trigger circuit' as it is very similar, as a circuit shape, to a very early circuit idea published by O. H. Schmitt [8] in 1938.

So, it does appear that positive feedback is a good idea for comparator circuits. Positive feedback can provide some kind of trigger threshold for the circuit to operate on when the input voltage passes some predetermined reference level, and can also provide hysteresis which, as Fig. 4.8 shows, implies that there can be a small region around the critical threshold within which the circuit will ignore fluctuations and noise on the input signal. These are useful features but positive feedback along the lines of Fig. 4.6 does not lead to a good comparator circuit. A far more subtle approach is called for which maintains the balanced input of the circuit shown in Fig. 4.4, and also takes the system considerations, discussed at length in section 4.3, into account, particularly the fact that a clock signal could be introduced into the comparator circuit.

4.6 Positive feedback and symmetry

The trouble with Fig. 4.6, as a circuit idea, is that the symmetry of the operational amplifier's input circuit, illustrated by Fig. 4.4, has been lost. One input is used for the signal and the other is used for the positive feedback. Is it possible to apply positive feedback to both input terminals? Not directly, and not through potential dividers, because this would virtually destroy the possibility of using these input terminals for a signal input. For example, a direct connection of positive output terminal to negative input terminal, and also negative output to positive input, in Fig. 4.4, simply makes the input and output terminals one and the same.

What is needed is an extra pair of input terminals, which may be used to accept the positive feedback, while the original input terminals continue to accept the signal. This idea is shown in Fig. 4.9, where two long tailed pair circuits have been wired in parallel, as far as their collector loads are concerned. Both have independent tail currents, I_{L1} and I_{L2}, and one pair, Q_1 and Q_2, accepts the input signal, while the other pair, Q_3 and Q_4, has 100% positive feedback applied to it.

The circuit of Fig. 4.9 has some very remarkable properties. If I_{L2} is made zero, the circuit is identical to Fig. 4.4 and is a simple amplifier with the transfer function shown in Fig. 4.5. If I_{L2} is now increased gradually from zero, positive feedback is being applied and the gain of the circuit increases until, when $I_{L2} R_L = 2V_T$, using the symbolism of section 4.4, the gain becomes infinite. Further increase of I_{L2} causes the circuit to become a trigger circuit, with the kind of transfer function shown in Figs. 4.7 and 4.8.

A full analysis of this kind of circuit will be given later in this chapter when the final form, which will be the experimental circuit for this chapter,

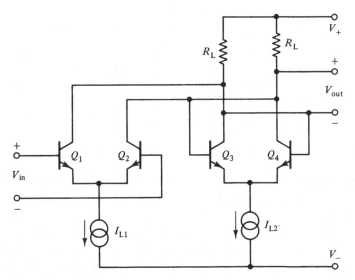

Fig. 4.9. *Applying positive feedback and yet maintaining a balanced circuit.*

has been reached. There are still a number of points to consider. What has been achieved so far is a circuit which can have radically different behaviour, illustrated by the contrast between the transfer functions shown in Figs. 4.5 and 4.8, simply by changing the value of a current, I_{L2}. This immediately suggests the possibility of using the clock signal to control I_{L2}. In fact, the key to an excellent comparator, which will really fit into a digital system, lies in the clock control of both I_{L1} and I_{L2}.

4.7 Introducing the clock signal

A circuit of the shape shown in Fig. 4.9 may have been first described by Lynes and Waaben [9] when it was applied to a sense amplifier problem. The use of the circuit as a comparator was presented by Slemmer [10] at about the same time. All these authors realised that a very important step forward in the technique of making very fast and very sensitive amplifying circuits, specifically within the environment of a digital system, could be taken if the clock signal was made to operate the circuit of Fig. 4.9 in two distinct states.

During what might be called the sense state, $I_{L1} = I_L$ and $I_{L2} = 0$. The circuit then operates as a straightforward amplifier, and a small output voltage, v_o, is developed. This will appear across the collector junction of Q_3 as $-v_o$, while across the collector junction of Q_4 it will appear as $+v_o$; Q_3 and Q_4 are off at the moment, and v_o is typically a few millivolts.

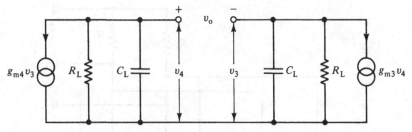

Fig. 4.10. An equivalent circuit for Fig. 4.9 when $I_{L1} = 0$ and $I_{L2} = I_L$.

During what might be called the latch state of the circuit shown in Fig. 4.9, $I_{L1} = 0$ and $I_{L2} = I_L$. Also $I_L R_L$ is made large enough to put Q_3 and Q_4 into regeneration, that is to have the behaviour shown in Fig. 4.8, and the final state which the circuit latches into will be decided entirely by the initial values of the collector junction voltages discussed in the previous paragraph. It is possible to take this discussion considerably further at this stage, because the equation of motion for this circuit is easily written.

Fig. 4.10 shows a very much simplified small signal equivalent circuit for the circuit shown in Fig. 4.9. It is assumed that $I_{L1} = 0$ and $I_{L2} = I_L$. The transistor Q_3 has v_3 across it, while the transistor Q_4 has v_4. The output voltage is then $v_o = v_4 - v_3$. The total capacitance from each output terminal to the common emitters of Q_3 and Q_4 is represented by C_L, and other capacitances are neglected. This is an accurate enough model to illustrate the important properties of this circuit. A full analysis has been given by Zojer, Petschacher and Luschnig [11].

The transistors Q_3 and Q_4 share the current $I_{L2} = I_L$ equally, so that their mutual conductances are

$$g_{m4} = g_{m3} = I_L/2V_T \qquad (4.8)$$

where the symbolism of equation (4.2) is being used again. The nodal equations for Fig. 4.10 are

$$g_{m4} v_3 + v_4/R_L + C_L \, dv_4/dt = 0 \qquad (4.9)$$

and

$$g_{m3} v_4 + v_3/R_L + C_L \, dv_3/dt = 0 \qquad (4.10)$$

As $v_o = v_4 - v_3$ it follows, from subtracting equation (4.10) from equation (4.9), that

$$C_L R_L \, dv_o/dt + (1 - g_m R_L) v_o = 0 \qquad (4.11)$$

where $g_m = g_{m3} = g_{m4}$, from equation (4.8).

Equation (4.11) may be solved by substituting $v_o = V \exp(at)$, where

V is the initial value of v_o that has been set by the input signal during the sense state of the circuit which was described above. This shows that

$$a = (g_m R_L - 1)/C_L R_L \qquad (4.12)$$

which shows that the circuit is regenerative once I_L is increased sufficiently to make $g_m R_L > 1$. From equation (4.8) this means that I_L must exceed $2V_T/R_L$, as stated above. The output voltage, v_o, would then increase exponentially, with the time constant $C_L R_L/(g_m R_L - 1)$ (see equation (4.12)) until the circuit ran into the non-linear regime and latched into a state where either Q_3 was on and Q_4 off, or *vice versa*.

Clearly, the circuit designer's goal would be to make the time constant $C_L R_L/(g_m R_L - 1)$ as small as possible. This can be done by increasing R_L, but once R_L exceeds $1/g_m$ by a factor of about 10, there is no point in any further increase because the time constant has fallen to a limit of C_L/g_m. In any case, R_L would not be made so large that the transistors Q_3 and Q_4 could saturate.

The circuit's time constant may, therefore, be taken to be C_L/g_m, and this could be well below 100 ps in an integrated circuit. It follows that Fig. 4.9 is a circuit shape of great potential, and should lead to an excellent design for a fast comparator circuit which will fit into a digital system environment. The final form of such a comparator circuit will now be given, and then experimental work can begin.

4.8 The final circuit shape

Fig. 4.11 shows the additional features that should be added to Fig. 4.9 in order to make the basis for an excellent fast comparator.

First, the load resistors, R_L in Fig. 4.9, have been split into two, R_1 and R_2 in Fig. 4.11. This idea is found in the work of Peetz, Hamilton and Kang [12], and is an excellent way of optimising the bandwidth of the sense amplifier state, which calls for a low value of R_L, and the speed of the latch state, which calls for a high value of R_L.

The second addition shown in Fig. 4.11 is that emitter followers, Q_5 and Q_6, have been added to reduce the capacitive loading, C_L in equation (4.11), and also to increase the reverse bias across the collector junctions of Q_3 and Q_4 by about 700 mV. This reduces C_L still further, and generally improves the gain-bandwidth of the circuit. The output from the circuit can then be taken from the emitters of Q_5 and Q_6.

Finally, Q_7 and Q_8 have been added to switch the I_L either into the sense amplifier, Q_1 and Q_2, or into the latch, Q_3, Q_4, Q_5 and Q_6, depending upon whether the clock signal is low or high.

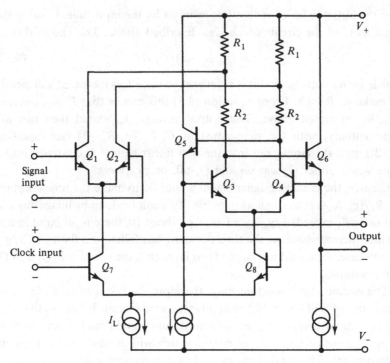

Fig. 4.11. *The final circuit shape for a high speed comparator circuit.*

Further additions would have to be made to the circuit in order to put the correct level shifts onto the inputs and outputs, and to make both the clock input accept, and the circuit output supply, signal levels compatible with the system's logic. All this would have to be done within the environment of one VLSI circuit, one which would include all the features shown in Fig. 4.2. Some examples of this kind of advanced design may be seen in the references that have already been given ([3], [4], [11] and [12]) and in the work of Inoue *et al.* [13], van de Grift and van de Plassche [14], and Yoshii, Asano, Nakamura, and Yamada [15]. A very interesting design using a new bipolar process has been published by Hotta *et al.* [16], and the circuit idea of Fig. 4.9 has been tried out in gallium arsenide [17] and in CMOS [18]. The problems of actually using these advanced circuits have been discussed in a very useful paper by Emmens and Lonsbourough [19].

4.9 An experimental circuit

The experimental circuits for this chapter will not be the advanced integrated circuits discussed above. In order to get an understanding of the interesting comparator idea shown in Fig. 4.11, it is only necessary to construct a minimum amount of hardware, and the need for reasonable device matching can be met by using transistor arrays.

The first experimental circuit is shown in Fig. 4.12 and may be used to investigate the quasi-static behaviour of the circuit: its critical signal levels and its hysteresis. This will illustrate some unexpected and important behaviour.

Fig. 4.12 is realised using two CA3046 transistor arrays. Q_1 and Q_2 belong to one array, while $Q_{1'}$ to $Q_{4'}$ belong to the other. The tail current for Q_1 and Q_2 has been made about 1 mA, and is kept constant during the experimental work. It is the tail current for the regenerative part of the circuit, that is I_{L2} in the previous Fig. 4.9, which can be varied by means of a variable supply, V_{L2} in Fig. 4.12, that is connected to R_8. This simulates the conditions under which the circuit would operate at high speed, because a constant tail current for Q_1 and Q_2 maintains the signal level across R_3 and R_5, which this sensing part of the circuit transfers to the latch part of the circuit. Such a situation would apply during fast clocking of the real circuit (Fig. 4.11) in which the signal level, referred to above, would remain stored on the collector junction capacitances of $Q_{1'}$ and $Q_{2'}$.

It was shown in section 4.7 that there was no point in increasing the load resistance for the latch part of the circuit above $10/g_m$. For the current levels chosen for Fig. 4.12, this means that $(R_3 + R_4)$, and $(R_5 + R_6)$, would be kept below 250 Ω. A value of 122 Ω has been chosen. This effects a sensible split in the load resistors, as discussed at the beginning of section 4.8, with $R_3 = R_5 = 22$ Ω and $R_4 = R_6 = 100$ Ω.

4.10 Hysteresis measurements

For the first experimental measurements, the circuit of Fig. 4.12 should be connected to a d.c. signal source that will provide an input of ± 50 mV. This can be done by using a variable power supply and a potential divider. A similar variable power supply should be used for V_{L2}. An oscilloscope on XY display should then be used to show the emitter voltage of $Q_{3'}$, or $Q_{4'}$, against V_{L2}.

Now set $V_{L2} = +5$ V and V_{in} to, say, $+10$ mV. This positive input signal would be expected to make the emitter of $Q_{4'}$ go low once the tail current of $Q_{1'}$ and $Q_{2'}$ is increased by reducing V_{L2}. It might be thought that this would happen discontinuously, at some critical level of V_{L2}, but this is not so. Referring to Figs. 4.5, 4.7 and 4.8, where the idea of circuit hysteresis was first introduced, it is clear that a positive input must produce a

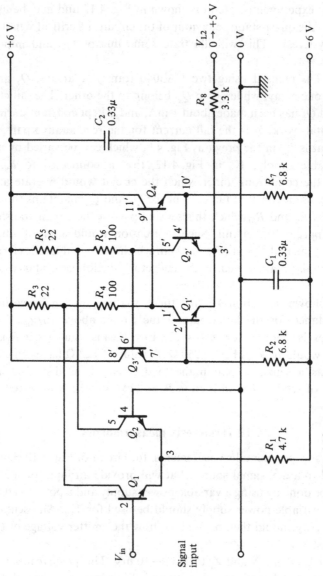

Fig. 4.12. An experimental circuit for static measurements on the comparator.

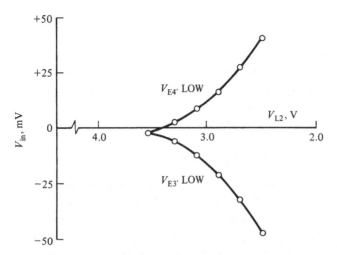

Fig. 4.13. *Experimental measurements of the hysteresis in the circuit shown in Fig. 4.12. V_{L2} is held constant and V_{in} varied over the range ± 50 mV. For increasing V_{in}, the discontinuous transition $V_{E3'}$ LOW to $V_{E3'}$ HIGH, and, of course, $V_{E4'}$ HIGH to $V_{E4'}$ LOW, occurs when the upper curve is crossed. The discontinuous transition, $V_{E4'}$ LOW to $V_{E4'}$ HIGH, and, of course, $V_{E3'}$ HIGH to $V_{E3'}$ LOW, occurs when the lower curve is crossed, during a decrease in V_{in}. The small negative offset input voltage, shown by these results, can be attributed to the particular CA3046 being used for Q_1 and Q_2.*

positive output when the *initial* state of the circuit has a transfer function of the kind shown in Fig. 4.5. Hysteresis only applies when the circuit has been put into the state represented by the transfer functions shown in Figs. 4.7 and 4.8, and will then only be observed if the input voltage is varied.

It follows that hysteresis may be observed if V_{L2} is reduced, step by step, and, at each step, V_{in} is varied slowly from, say, -50 mV to $+50$ mV, and then back again. As V_{in} is varied, a sudden discontinuous change in voltage level at the emitter of $Q_{3'}$, and $Q_{4'}$, should be observed. The values of V_{in} and V_{L2} at which these discontinuities occur lie on the two curves shown in Fig. 4.13. It is the space between these two curves which is equal to the width of the hysteresis loop, shown in Fig. 4.8. As the tail current of $Q_{1'}$ and $Q_{2'}$ increases, so increasing the loop gain of this regenerative circuit, the hysteresis loop gets wider and wider. When $V_{L2} = 0$, the loop is so wide that the circuit is really 'latched': no matter how large the input signal, the voltage across R_3 and R_5, due to Q_1 and Q_2 is not big enough to push the circuit into its other state.

4.11 Hysteresis theory

Before asking what the implications of the circuit hysteresis might be, which, of course, brings up the dynamic performance, it is interesting to find out why the experimental results lie on two curves of the shape shown in Fig. 4.13.

Referring to Fig. 4.12, the difference between the currents in Q_1 and Q_2 is

$$I_{C1} - I_{C2} = I_{L1} \tanh (V_{in}/2V_T) \qquad (4.13)$$

where I_{L1} is the current in R_1. This causes a voltage $I_{C1} R_3 - I_{C2} R_5$ to be transferred to $Q_{1'}$ and $Q_{2'}$. Because $R_5 = R_3$ this voltage may be written $(I_{C1} - I_{C2}) R_3$. When V_{L2} falls from $+5$ V to zero, and I_{L2} begins to flow, an additional voltage $(I_{C1'} - I_{C2'}) (R_3 + R_4)$ will be added. This follows because $R_4 = R_6$. The action of the circuit may be concisely summed up by the single equation

$$y = x \tanh (y + z) \qquad (4.14)$$

where

$$y = (I_{C1'} - I_{C2'}) (R_3 + R_4)/2V_T \qquad (4.15)$$

$$x = I_{L2}(R_3 + R_4)/2V_T \qquad (4.16)$$

and

$$z = (I_{L1} R_3/2V_T) \tanh (V_{in}/2V_T). \qquad (4.17)$$

Now the experimental results of Fig. 4.13 were presented as measurements made by fixing V_{L2}, that is I_{L2}, and varying V_{in}. For analysis, however, it is simpler to consider the alternative, but more tedious, experiment in which V_{in} is fixed, and I_{L2} is decreased to take the representative point, (V_{L2}, V_{in}) in Fig. 4.13, across the critical hysteresis boundary. From this point of view, z, given by equation (4.17), is a constant, and the hysteresis boundary is given by the values of I_{L2} at which dy/dx becomes infinite.

From equation (4.14)

$$dy/dx = \tanh (y + z) + [x/\cosh^2 (y + z)] dy/dx \qquad (4.19)$$

which, using equation (4.14) to substitute for x and rearranging, may be written

$$dy/dx = [\tanh (y + z)]/[1 - 2y/\sinh 2(y + z)]. \qquad (4.20)$$

It follows that $dy/dx = \infty$ at values of y given by the solution of the transcendental equation

$$\sinh 2(y + z) - 2y = 0. \qquad (4.21)$$

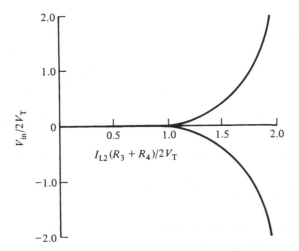

Fig. 4.14. *The theoretical hysteresis boundaries for the circuit shown in Fig. 4.12, calculated for* $I_{L1} R_3 / 2 V_T = 0.5$.

These values of y may then be used in equation (4.14) to find the corresponding values of x. All this is easily done on a programmable calculator. The program takes in a value of $V_{in}/2V_T$, calculates z from equation (4.17), finds the root of equation (4.21) by using the Newton–Raphson method, and finally gives the value of x using equation (4.14).

The results are shown in Fig. 4.14 and, when these are scaled correctly using the component values given in Fig. 4.12, very close agreement is found between these theoretical curves and the experimental results shown in Fig. 4.13, for small values of V_{in}. The experimental circuit does not show a clear saturation for large positive values of V_{in} because a simple resistor, R_1 in Fig. 4.12, has been used to define I_{L1}, instead of a constant current sink. This means that Q_1 has a gain of R_3/R_1 to large positive inputs, instead of giving the zero gain shown in Fig. 4.5.

4.12 The implications of hysteresis

Hysteresis, of the kind considered in the previous two sections, can give rise to interesting time dependent errors in digital systems which dispense with a sample and hold circuit in advance of the ADC, and call for the ADC to implement its own sample and hold function under the control of the clock signal. This problem has been discussed very clearly by Peetz *et al.* [12]. The clock signal will be arranged to switch the circuit from the sense state to the latch state, as described in section 4.7, as rapidly as possible. The effect of this may be imagined by considering the transfer functions shown in Figs. 4.5, 4.7 and 4.8. In a matter of a few nanoseconds the transfer function becomes a hysteresis loop, and then the width of this hysteresis loop begins to expand, starting with negligible width. The actual value of the width, in terms of input voltage variation, is given by the distance between the two curves shown in Fig. 4.14.

Once inside the hysteresis loop, the representative point is captured unless it is moving so fast that it can catch up and overtake the other side of the expanding loop. This can clearly happen with very fast signals that enter the loop when it has only just started to expand. As Fig. 4.14 shows, once I_{L2} is really established, the velocity of hysteresis loop expansion becomes very large indeed.

If the representative point does succeed in crossing the hysteresis loop before it fully expands, then the circuit will latch high for the signal coming from the $V_{in} < 0$ side, and low for one entering from the $V_{in} > 0$ side. This is the opposite way around when compared to slow signals, despite the fact that the actual value of both fast and slow signals was the same at the instant the clock signal initiated the latch state of the circuit.

It is for these reasons that a sample and hold circuit must precede the ADC in a very fast digitising oscilloscope. When lower frequencies are being handled, it is possible to make the ADC operate as its own sample and hold [12].

4.13 Dynamic measurements

Fig. 4.15 shows a second experimental circuit. This can be used to make dynamic measurements on the high speed comparator circuit idea that was first shown in Fig. 4.11.

The circuit shown in Fig. 4.15 is easily constructed by adding some more work to the first experimental circuit, the one shown in Fig. 4.12. The transistors Q_3 and Q_4, in Fig. 4.15, are already available on the CA3046 which contains Q_1 and Q_2. Other components have been added to make Q_3 and Q_4 perform the same function that Q_7 and Q_8 perform in Fig. 4.11: a constant tail current is switched repetitively either to Q_1 and Q_2, for the sense state of the circuit, or to $Q_{1'}$ and $Q_{2'}$, for the latch state.

The clock signal for the circuit shown in Fig. 4.15 is provided by a simple square wave generator which is RC coupled to the base of Q_4. This makes for a very simple experimental circuit, but care must be taken not to overdrive Q_4. To set the correct square wave input level, which will be less than 150 mV peak to peak, the signal at the top of R_1, in Fig. 4.15, should be viewed on a d.c. oscilloscope. A square wave, centred on a d.c.

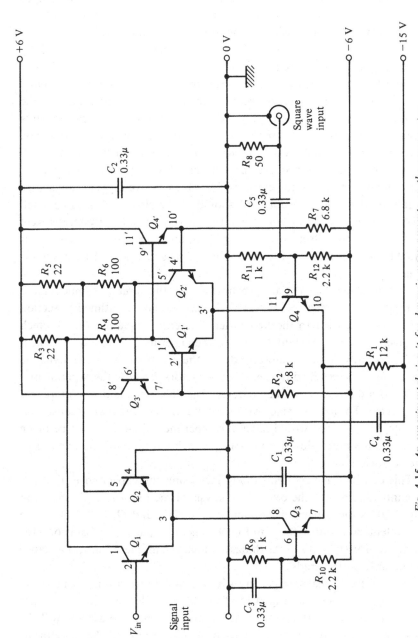

Fig. 4.15. An experimental circuit for dynamic measurements on the comparator.

level between -2.5 V and -3.0 V, should be seen, and the correct input level will have been reached just before the negative amplitude of this square wave stops increasing as the input level is increased.

The dynamic performance of the circuit may now be observed by connecting two high impedance probes to the emitters of $Q_{3'}$ and $Q_{4'}$ and viewing the waveforms on a dual beam oscilloscope, using a.c. coupling and 50 mV/div sensitivity. A small d.c. input signal is then applied to the circuit, in just the same way as was done in the hysteresis measurements described in section 4.10, and this d.c. input signal is varied from, say, -25 mV to $+25$ mV. A square wave of about 100 mV peak to peak should then be observed on the emitter of $Q_{3'}$, for negative inputs, changing over to be on the emitter of $Q_{4'}$, for positive inputs. The change over should be very abrupt, and should occur close to the same small offset voltage that was found when the measurements of hysteresis were made. This is the offset shown in Fig. 4.13, but note that exact agreement should not be expected, for an important reason that will be discussed later.

At this stage, the dynamic behaviour of the circuit is much more complicated than that predicted by the simple model outlined in section 4.7. This model led to the first order differential equation (4.11) which involved the time constant C_L/g_m. In the experimental circuit, Fig. 4.15, C_L/g_m is only a few nanoseconds at the moment, and the dynamic behaviour is determined by a number of factors, because the experimental circuit is a discrete component circuit which has been laid out on quite a large scale. To make contact with the kind of dynamic performance that an integrated circuit would have, the experimental circuit must be made to work on a much slower time scale, as with the experimental circuits of previous chapters.

This can be done by deliberately increasing the capacitance, C_L, that was introduced into the dynamic analysis of section 4.7. C_L is shown in Fig. 4.10 as the capacitance which is across Q_3 and Q_4 in Fig. 4.9. The equivalent capacitance to C_L in Fig. 4.15 is the capacitance from collector to emitter for $Q_{1'}$ and $Q_{2'}$. As things stand at the moment, this capacitance is only a few picofarads.

If C_L is made really large, say 470 pF, by connecting two capacitors across $Q_{1'}$ and $Q_{2'}$, in Fig. 4.15, from their collectors down to their common emitters, the most interesting dynamic behaviour will be observed that corresponds to the kind of behaviour, and the waveforms, published by Zojer *et al.* [11] for an advanced integrated circuit. The time scale of the experimental circuit is, of course, much slower.

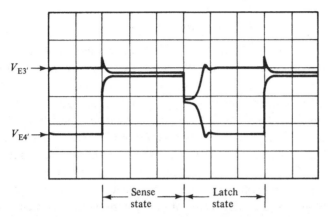

Fig. 4.16. *The waveforms observed at the emitters of* $Q_{3'}$ *and* $Q_{4'}$, *in Fig.* 4.15, *when there is a positive input voltage, and when both* $Q_{1'}$ *and* $Q_{2'}$ *have* 470 pF *capacitors connected from their collectors down to their common emitters. The display is a.c. coupled,* 50 mV/div *and* 1 μs/div. *The two waveforms are shown with their correct relative d.c. levels.*

Fig. 4.16 shows the waveforms which should be observed at the emitters of $Q_{3'}$ and $Q_{4'}$, in Fig. 4.15, when V_{in} is just above the critical threshold level, that is for a small positive value of V_{in}. The two waveforms have been adjusted to lie at the correct relative d.c. levels.

During the sense state, Fig. 4.16 shows that the emitters of $Q_{3'}$ and $Q_{4'}$ each have a small signal level on them, $V_{E4'}$ being slightly more negative, relative to $V_{E3'}$, because of the small positive input signal to Q_1 and Q_2. This difference in level between $V_{E4'}$ and $V_{E3'}$ has been exaggerated slightly in Fig. 4.16, for clarity, but the difference will be seen very clearly if V_{in} is increased to, say, +25 mV.

The latch state begins when the square wave input turns Q_4 on and Q_3 off. The first thing that happens is that the emitters of $Q_{3'}$ and $Q_{4'}$ both fall together by an amount $I_{L2}(R_3 + R_4)/2$. This follows because half of the current in Q_4 now flows in $(R_3 + R_4)$ and the other half in $(R_5 + R_6)$. The drop is, therefore, about 50 mV.

The circuit is now regenerative, in the sense of equation (4.11) with $g_m R_L > 1$. The *difference* between the two voltages displayed in Fig. 4.16 now grows exponentially until the circuit leaves the simple linear regime, used in the analysis leading to equation (4.11), and takes on the levels determined by the state in which $Q_{1'}$ is on and $Q_{2'}$ is completely switched off.

4.14 The dynamic offset voltage

A very important point about the dynamics of this circuit should have
been revealed by these experiments. When the two 470 pF capacitors were
connected across $Q_{1'}$ and $Q_{2'}$, in Fig. 4.15, a large change in the offset
voltage, the critical input level at which $V_{E3'}$, or $V_{E4'}$, changes from a large
square wave to a small one, should have been observed. The offset voltage
should now be quite large, and may be positive or negative. The reason for
this is the tolerance on the two capacitors used, and this may be confirmed
very easily by simply interchanging them: the large offset voltage will
change sign. This observation is important because it explains why the
offset voltage observed in the earlier static hysteresis measurements, the
offset shown in Fig. 4.13, differs from the offset observed in the dynamic
tests, described in section 4.13: the difference is due to the different stray
capacitances associated with the layout of this discrete component circuit.

Why should a difference in the capacitance across $Q_{1'}$ and $Q_{2'}$, in Fig.
4.15, cause a change in d.c. offset voltage, which is surely due to the simple
d.c. offset voltage of the input pair, Q_1 and Q_2? The answer to this may
be seen in the waveforms shown in Fig. 4.16. At the beginning of the latch
state, Q_4 turns on and the current in Q_4 should be shared equally by $Q_{1'}$
and $Q_{2'}$. That is why the collector voltages of both $Q_{1'}$ and $Q_{2'}$ should drop
together, as shown in Fig. 4.16, and why the small difference between
these two voltages, initially set by the signal input, should be maintained.
Q_4 takes a finite time to turn on, however, and if the stray capacitances in
parallel with $Q_{1'}$ and $Q_{2'}$ are not identical, these capacitances will charge
up a small amount, during the finite turn-on time, but they will not charge
up equally, even if the current in Q_4 is shared equally between them.

But this brings out a further question. Why should the current in Q_4 be
shared equally between $Q_{1'}$ and $Q_{2'}$? For this to happen, $Q_{1'}$ and $Q_{2'}$ must
not only have identical d.c. characteristics, they must also be identical
dynamically. This calls for identical capacitance to be associated with the
emitter and collector junctions of both transistors, which implies perfect
symmetry in the layout of all parts of the circuit shown in Fig. 4.15 that
are associated with $Q_{1'}$ and $Q_{2'}$. This is, of course, impossible with a
discrete component circuit, but it is almost possible with the layout of an
integrated circuit, although this may not be as easy as one might think.
For example, it is obvious that the connections which must be made to the
collectors of Q_1 and Q_2, the input sense amplifier, will involve crossovers
which are asymmetric if the circuit is laid out as it is drawn in Fig. 4.15.
If this is avoided by placing Q_1 and Q_2 in between R_4 and R_6, it is now the
connections to the bases of Q_1 and Q_2 that will cause problems, to say

nothing of the connection to their emitters. When all these problems have been solved, the designer still has to deal with the layout of 255 such circuits, all on one die, and their associated logic, as shown in Fig. 4.2. It is, in fact, the problems associated with relative propagation delays [12], between various parts of such a large scale integrated circuit, that cause the greatest difficulty. Because the devices used are so fast, and the capacitances associated with individual devices are so small, the problem of unequal stray capacitance, which has been illustrated by the experimental circuit used here, is not the most difficult problem the designer has to solve.

4.15 Conclusions

This chapter began by considering the ADC problem in general and its importance in the fields of instrumentation, measurement, and communications. The comparator circuit was then isolated as a key circuit problem in ADC design, particularly for the flash ADC which is the fastest of all ADCs and the one used in the fastest digitizing oscilloscopes.

System considerations were then brought in to show how the idea for a new circuit shape could come from an understanding of what an entire system, of which the new circuit was to be only a minor, but essential, part, was intended to do. This led to a comparison between a simple operational amplifier kind of comparator and a comparator using positive feedback.

Positive feedback makes a comparator circuit have a critical input voltage, a threshold level, at which the output changes state. Positive feedback also introduces hysteresis into the transfer function of the comparator, which can be a very useful feature if it is properly controlled.

The idea of control led to a discussion of how the clock signal of the digital system would be used to control the transfer function of the comparator circuit so that this transfer function could take on quite distinct forms, the different forms that are required during different times in the data processing cycle. It is interesting to note that this idea, which led to the circuit shape shown in Fig. 4.11, did not appear until 1969 [9], and was not widely adopted until Peterson took it up in 1979 [4]. Digital systems using comparators have been around since the 1940s, however [2].

The experimental circuit for this chapter made it possible to measure the hysteresis boundaries of the circuit, under really static conditions, and to compare these measurements with theory. Hysteresis, in the fast comparator/latch circuit, has interesting implications when the circuit must handle signals that are changing very rapidly with time. This is very important in systems which dispense with a true sample and hold circuit, the kind of circuit discussed in chapter 3, and make direct use of the comparator/latch to measure the input signal amplitude [12] at some instant in time.

Hysteresis is only one of the problems facing the circuit designer when the dynamics of the comparator/latch circuit are considered, and this chapter closed with a description of some dynamic experimental work that may be done by adding some more hardware to the first experimental circuit. There is much that may be done with this final circuit, Fig. 4.15, both with and without the additional capacitors that should be added in order to slow the circuit down so that it behaves in a fairly simple way. A very clear demonstration of the statistical nature of the circuit's response to d.c. signals very close to the threshold input voltage may be made, and it is also interesting to examine the circuit's response to a small sinusoidal input. Another possibility is to increase the values of R_4 and R_6 so that the circuit may be studied under conditions which allow $Q_{1'}$ and $Q_{2'}$ to saturate.

Notes

1 Gordon, B. M., *IEEE Trans. Circ. Syst.*, **CAS-25**, 391–418, 1978.
2 Kandiah, K., *Nucl. Instr. Meth.*, **162**, 699–718, 1979.
3 Frohring, B. J., Peetz, B. E., Unkrich, M. A., and Bird, S. C., *Hewlett–Packard J.*, **39**, No. 1, 39–47, Jan. 1988.
4 Peterson, J. G., *IEEE J. Sol. St. Circ.*, **SC-14**, 932–7, 1979.
5 O'Dell, T. H., *Electronic Circuit Design, Art and Practice*, Cambridge University Press, Cambridge, 1988, chapter 8.
6 Horowitz, P., and Hill, W., *The Art of Electronics*, Cambridge University Press, Cambridge, second edition, 1989, pp. 231 and 580–2.
7 Williams, F. C., *J. IEE*, **93**, Part IIIA, No. 1, 289–308, 1946.
8 O. H. Schmitt published a number of very interesting electronic circuit ideas during a visit he made, from the USA, to University College, London, in 1938. These can all be found in Vol. 15 of the *Journal of Scientific Instruments*. He returned to the USA to become one of the principal workers in the field of medical electronics.
9 Lynes, D. J., and Waaben, S. G., US Pat. No. 3480800, 1969.
10 Slemmer, W. C., *IEEE J. Sol. St. Cir.*, **SC-5**, 215–20, 1970.
11 Zojer, B., Petschacher, R., and Luschnig, W. A., *IEEE J. Sol. St. Circ.*, **SC-20**, 780–5, 1985.
12 Peetz, B., Hamilton, B. D., and Kang, J., *IEEE J. Sol. St. Circ.*, **SC-21**, 997–1002, 1986.
13 Inoue, M., Sadamatsu, H., Matsuzawa, A., Kandu, A., and Takemoto, T., *IEEE J. Sol. St. Circ.*, **SC-19**, 837–41, 1984.
14 van de Grift, R. E. J., and van de Plassche, R. J., *IEEE J. Sol. St. Circ.*, **SC-19**, 374–8, 1984.
15 Yoshii, Y., Asano, K., Nakamura, M., and Yamada, C., *IEEE J. Sol. St. Circ.*, **SC-19**, 842–6, 1984.
16 Hotta, M., Shimizu, T., Maio, K., Nakazato, K., and Ueda, S., *IEEE J. Sol. St. Circ.*, **SC-22**, 939–43, 1987.

17 Ducourant, T., Baelde, J.-C., Binet, M., and Rocher, C., *IEEE J. Sol. St. Circ.*, **SC-21**, 453–6, 1986.
18 McCarroll, B. J., Sodini, C. G., and Lee, H.-S., *IEEE J. Sol. St. Circ.*, **SC-23**, 159–65, 1988.
19 Emmens, T., and Lonsbourough, M., *EDN*, **27**, No. 6, 137–43, 17 March, 1982.

5
Probes and input circuits

5.1 Introduction

An electronic instrument needs to be connected to whatever circuit or system is under test. Very wide bandwidth instruments, those used for signals above 100 MHz, usually have 50 Ω co-axial inputs. Instruments which accept signals of even higher frequency, in the microwave and optical range, will have waveguide or optical fibre inputs. The more conventional laboratory test equipment is usually supplied with some kind of probe input circuit which may be connected to the circuit under test.

The probes and input circuits used with oscilloscopes provide good examples of the techniques employed. These same probes and input circuits may, of course, be used in a variety of instruments: vector voltmeters, network analysers, spectrum analysers, and so on. Very high impedance probes are used for voltage measurement, while very low impedance sensor probes must be inserted for current measurement. These voltage and current probes may be passive or active, and both kinds will be considered in this chapter.

Voltage and current measurements are not the only ones that are called for in electronics. Measurements of incident, transmitted and reflected power may also be required. This is the approach often used for high frequency, wide-band circuits which work as part of a transmission line system; for example, a repeater amplifier in a cable television system. The input circuits needed for these power measurements are particularly interesting in that they may exhibit directional properties. Such directional circuits are considered at the end of this chapter.

To begin with, however, it is useful to look at the simplest of input circuits: the kind of circuit which lies behind the input socket of any instrument. It is at this point that the circuit designer must consider what

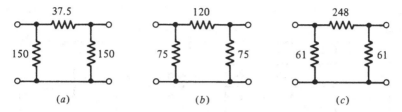

Fig. 5.1. *Showing the π networks that may be used to give attenuations of (a) × 2, (b) × 5 and (c) × 10, in a 50 Ω transmission line system.*

the user of an instrument may, unexpectedly, connect to this input. Some protection should always be provided. The circuit which lies behind the input socket also determines what kind of voltage or current probe may be used.

5.2 Input circuits for 50 Ω systems

Fig. 5.1 shows three well-known 50 Ω π networks which may be used to give the 2, 5, 10 sequence of attenuation steps that is so often found.

Such simple π networks can be made, using discrete resistors, which will prove quite satisfactory 50 Ω input circuits for oscilloscopes with bandwidths up to 250 MHz. The reason is that unwanted capacitance in the networks, due to the finite size of the resistors and the connections that must be made between them, will present an impedance much higher than 50 Ω. A capacitance of 10 pF, which is really big for a stray capacitance, has an impedance of 50 Ω only when the frequency is just over 300 MHz.

The three networks shown in Fig. 5.1 are all that are needed to build up a switched 50 Ω input attenuator for any instrument. For example, an oscilloscope with a sensitivity switched between 10 mV/div and 1 V/div in the 2, 5, 10 sequence. Fig. 5.2 shows how this may be done.

At frequencies above 250 MHz, the same circuits may be used, but with thin film resistors and with quite advanced switching technology to overcome the problems of stray capacitance which were mentioned above.

At really high frequencies, the instrument usually has a fixed sensitivity at its input socket. A reduction in input sensitivity is then best achieved by means of fixed attenuator pads. These can be built, as distributed circuits, to have a well-defined, constant attenuation over a bandwidth of several gigahertz, even when the attenuation is as high as 20 db per pad.

5.3 High impedance input circuits

When a wide-band instrument, like an oscilloscope, is designed to have a high input impedance, the problem of unwanted capacitance is much

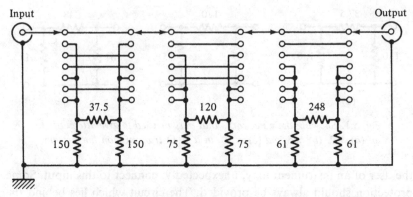

Fig. 5.2. *The three networks shown in Fig.* 5.1 *may be switched into combinations which give overall attenuations of* 0, × 2, × 5, × 10, × 50 *and* × 100.

more serious than in the case of 50 Ω input impedance instruments. It is virtually impossible to reduce the shunt capacitance across the input socket of the instrument below 20 pF. This capacitance is simply due to the input socket itself, the connection between this socket and the first stage of wide-band amplification, and then the input capacitance of the wide-band amplifier itself.

This means that, if the input impedance at low frequency is defined at 1 MΩ, which is a very common choice, the input impedance will begin to fall as $1/\omega$ once the input signal frequency exceeds $1/2\pi CR$, where $C = 20$ pF and $R = 1$ MΩ. This is a frequency of just over 7.5 kHz: not even beyond the audio range. It follows that a 1 MΩ, 20 pF, input point will present an impedance of only 800 Ω at 10 MHz, which is certainly not a very high frequency. In addition the oscilloscope must be connected to the circuit or system under test by means of a short length of cable, which, even if it is a really low capacitance cable, will have a capacitance of at least 50 pF/m. It is clear that something must be done to increase the impedance at the point where the oscilloscope is actually connected to the circuit under test.

5.4 Passive voltage probes

The simplest solution to the problem of low input impedance is the well-known 'high impedance probe' [1]. This is a passive device which increases the input impedance, at the actual point where the instrument is connected to the circuit under test, by the same factor with which it reduces the overall sensitivity. The high impedance probe is connected to the instrument by a convenient length of low capacitance cable.

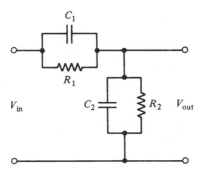

Fig. 5.3. *High impedance input circuits are usually based upon this all-pass network in which* $C_1 R_1 = C_2 R_2$.

For oscilloscopes, and other instruments, which can handle frequencies up to 10 MHz, the cable which connects the high impedance probe to the instrument may be treated as a lumped capacitance. This assumption allows the action of the high impedance probe to be understood by considering the all-pass network shown in Fig. 5.3. The transfer function of this is

$$V_{out}/V_{in} = R_2/[R_2 + R_1(1 + j\omega C_2 R_2)/(1 + j\omega C_1 R_1)] \qquad (5.1)$$

which shows that the attenuation may be made equal to $R_2/(R_2 + R_1)$ at all frequencies, provided the two time constants, $C_1 R_1$ and $C_2 R_2$, are made equal. The idea is applied by making $C_2 R_2$ the input impedance of the oscilloscope itself, which means that $R_2 = 1$ MΩ and $C_2 = 20$ pF, plus the capacitance of the cable which connects the probe to the oscilloscope: this can be 50 pF, making C_2 total at 70 pF. The components $C_1 R_1$ are then connected to the input end of the cable to form the probe tip which is actually connected to the circuit under test. A $\times 10$ probe, for example, would call for $R_1 = 9$ MΩ and $C_1 = 7.77$ pF, assuming the value of 70 pF for C_2 which was discussed above. Some way of adjusting either C_1 or C_2 would have to be arranged if the probe were to be used on different oscilloscopes.

This very simple model for the high impedance probe is quite inadequate at frequencies well above 10 MHz when the cable connecting the probe to the oscilloscope becomes an appreciable fraction of a wavelength. The design of high impedance probes for really wide bandwidth oscilloscopes has been dealt with in an important paper by McGovern [2], who deals with the fact that most commercially available high impedance probes use a special cable that has a high resistance inner conductor. McGovern gives a number of useful references.

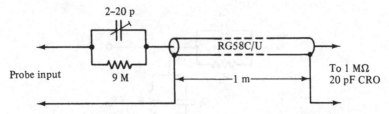

Fig. 5.4. *The circuit for an experimental* × 10, 10 *MΩ, passive probe.*

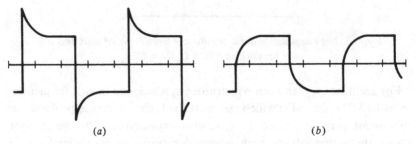

(a)   (b)

Fig. 5.5. *When the probe shown in Fig. 5.4 is connected to a low frequency square wave, from a low impedance source, the waveform (a) is observed when the 2–20 pF trimmer is adjusted to too large a value, and the waveform (b) is observed when the trimmer is adjusted to too small a value. When adjusted correctly, the probe reproduces the square wave perfectly, attenuating all frequency components by a factor of 10.*

5.5 An experimental high impedance probe

Even at frequencies below 10 MHz, the problem of designing a high impedance probe is not quite so simple as equation (5.1) might suggest, and it is worthwhile doing an experiment to show this. Fig. 5.4 shows what is needed. Take a 1 m length of ordinary 50 Ω co-axial cable with BNC connectors at either end. Because this cable will have a capacitance of, say, 100 pF, the input end will look like 120 pF in parallel with 1 MΩ when the cable is connected to a 1 MΩ, 20 pF, oscilloscope input.

Take a BNC panel socket, bolt a solder tag on one of the four holes and solder on a stiff 2 cm length of copper wire to act as a ground probe. Solder the 9 MΩ and the 2–20 pF trimmer capacitor directly on to the centre conductor of the socket, but note that the side of the trimmer which makes connection to the adjusting screw should be on the input side, not the socket side.

When this probe circuit is connected to a square wave, from a low impedance source and at a frequency of about 1 kHz, the waveforms shown in Figs. 5.5(a) and (b) will be observed, depending upon whether the trimmer capacitor is adjusted above or below the critical value which

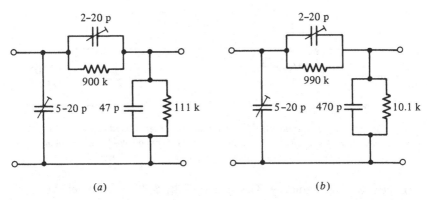

Fig. 5.6. *The input attenuators typically found inside an analog oscilloscope with a* 1 *MΩ input impedance. Network* (a) *gives* × 10 *attenuation and network* (b) *gives* × 100. *The* × 2 *and* × 5 *factors, which are also needed, would be provided by switching the feedback networks associated with the input amplifier.*

makes $C_1 R_1 = C_2 R_2$. The time constant involved, shown by the exponential rise or fall in Fig. 5.5, is around 120 μs because this is the order of both $C_1 R_1$ and $C_2 R_2$.

However, this simple test circuit will never be made to perform like a high impedance probe of the kind that is supplied with a commercial analog oscilloscope. There will always be a small error in the high frequency attenuation when the probe circuit is adjusted to give a good low frequency square wave response. The reason for this lies in the kind of input attenuator circuits found inside a typical analog oscilloscope with a 1 MΩ input impedance. These are shown in Fig. 5.6.

5.6 The problem of input capacitance

Fig. 5.6 shows that the same all-pass network idea, first shown in Fig. 5.3, is again being used in the input attenuator of the analog oscilloscope with a 1 MΩ input impedance. The 2–20 pF trimmer capacitor, shown in both Figs. 5.6(a) and (b), is adjusted to give a perfect low frequency response when the instrument is connected to a low impedance source.

Both Figs. 5.6(a) and (b) show a 5–20 pF trimmer capacitor connected across the input. The value of this trimmer is quite irrelevant when the input is connected to a low impedance source, and the reason it is there is that it can be adjusted to make the input capacitance of the instrument itself a constant, regardless of the setting of the internal attenuator. Unfortunately, this means that it is very difficult to get a high impedance probe of the kind shown in Fig. 5.4, one in which C_1 is a variable, to work

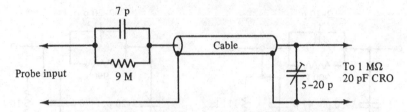

Fig. 5.7. *When the adjustable capacitor in a high impedance probe is connected across the oscilloscope end of the probe cable, it is much easier to get really good high frequency performance.*

correctly at high frequency. The setting of the 5–20 pF trimmer shown in both Figs. 5.6(*a*) and (*b*) may not be correct for both conditions: constant instrument input capacitance *and* all-pass properties for the combination of probe and input attenuator.

This is why nearly all high impedance probes adopt a circuit like the one shown in Fig. 5.7, in which the input components, 7 pF and 9 MΩ in this particular case, are fixed, and the probe adjustment is made by means of a trimmer capacitor at the oscilloscope end of the probe cable. It is then possible to get the correct × 10 probe attenuation at both low and high frequency. There is a further advantage in that the trimmer capacitor shown in Fig. 5.7 has one end grounded and is also located in a much more convenient place.

5.7 Errors caused by passive voltage probes

Errors of adjustment, discussed in the previous section, are not the only errors to be cautious of when using high impedance voltage probes on analog oscilloscopes. Whenever the probe is connected to a finite source impedance, there may be very serious errors introduced.

For example, suppose a × 10 probe has an input impedance of 10 MΩ in parallel with about 7 pF, as would be the case for the arrangement shown in Fig. 5.7. Just as in the case discussed in section 5.3, this 10 MΩ input impedance begins to fall as $1/\omega$ when the signal frequency is only in the kilohertz region. When the frequency is 10 MHz, this 'high impedance probe' looks like about 2.5 kΩ. This may cause considerable errors in amplitude, phase and rise time measurements [3]. It is fairly safe to say that measurements made with oscilloscopes using 10 MΩ, × 10, probes are accurate only at frequencies below 10 MHz, and then only when the source impedance being probed is below 100 Ω. As 10 MHz is by no means a high frequency in today's electronics, this problem is one which must be borne in mind.

Very wide bandwidth oscilloscopes, both analog and digitising, usually have only a 50 Ω input impedance, although a 1 MΩ input facility is often provided so that these very wide bandwidth instruments may also be used to make simple low frequency measurements. When a 50 Ω input impedance is unsuitable for some measurement which needs to be made at the really high frequencies, then probing is best done with an *active* voltage probe. These are described in the next two sections.

5.8 Classical active voltage probes

An active voltage probe is one, as the name implies, which uses active devices to provide a high impedance at the probe tip. There are two distinct kinds of active probe. The classical variety uses a microcircuit actually inside the probe tip, which is then connected, via a cable which must carry both signal and power supply lines, to some kind of connector box which presents the signal to the oscilloscope, and also picks up the power supply needed. A more modern kind of active probe is one which puts all the active devices into the connector box so that the probe tip can be made extremely small. Both kinds of active probe use interesting circuit shapes.

Fig. 5.8 shows an example of the kind of circuit shape found in the classical variety of active voltage probe [4]. This active probe is intended to provide an overall voltage gain of unity, a bandwidth of several hundred megahertz, and to be used with a very wide bandwidth oscilloscope having an input impedance of 50 Ω.

The probe tip, shown in Fig. 5.8, contains an amplifier which must provide the power gain needed for a voltage gain of unity and an impedance transformation from R_1, the high input impedance of the probe, down to the 50 Ω input impedance of the oscilloscope. This amplification is done with Q_1, a junction FET connected as a source follower, followed by a microcircuit with a voltage gain of 2, which can be matched, by means of R_4, to the 50 Ω cable that carries the signal on to the oscilloscope.

Note that this probe tip amplifier only handles the a.c. part of the input signal. The time constant $C_1 R_2$ decides the low frequency cut-off point for the probe tip, and would be chosen to lie in the kilohertz region. With $R_2 = 10$ MΩ, C_1 need then be only 100 pF, and this would mean that all the components associated with the gate of Q_1 would be of very small size and the probe tip input capacitance could be made very small indeed: only 2–3 pF.

The d.c. part of the input signal is dealt with at the far end of the probe

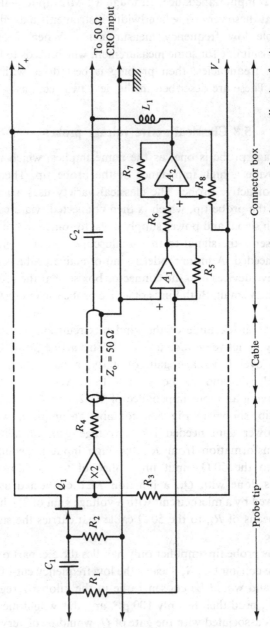

Fig. 5.8. *An active voltage probe which has a microcircuit in the probe tip and an overall gain of unity.*

cable, inside the connector box. This allows the probe tip to be made smaller because there are less components inside it. The d.c. signal path is simply an amplifier, A_1 with feedback components, $R_1 = R_5$, to give unity gain. The real part of the probe input impedance is R_1, and this is made 100 kΩ.

Because A_1 is an inverting amplifier, a second inversion is needed before the d.c. signal can be passed to the oscilloscope input. This is done by means of A_2, with feedback components $R_6 = R_7$. This second amplifier can also be used to introduce a d.c. offset facility, and this is shown as R_8.

An interesting general point should be made about the relative frequency responses of the a.c. and d.c. paths. The a.c. path, through Q_1, the $\times 2$ amplifier, and then the cable, *via* C_2, into the oscilloscope, will have its gain rising at 12 db/octave if $C_1 R_2$ and $C_2 \times 50$ Ω are made equal to one another. As stated above, the low frequency cut-off point would be made in the kilohertz region. The d.c. signal path, R_1 through A_1 and then A_2, only needs to have unity gain up to a frequency marginally above the low frequency cut-off point of the a.c. path. Because the a.c. and d.c. signal paths are in *parallel*, it does not then matter how the gain of this d.c. path falls as the frequency increases.

Finally, a very small inductor, L_1, is added to ensure that the output impedance of A_2 does not load the high frequency signal circuit in the megahertz region.

5.9 Recent developments in active probes

The most recent digitising oscilloscopes have effective bandwidths out in the gigahertz region, and, of course, have 50 Ω input impedance. To make the equivalent of a high impedance probe for instruments with this kind of very wide bandwidth calls for some radically new circuit ideas and constructional techniques. The aim is to get the capacitance at the probe tip down to as small a value as possible.

Fig. 5.9 shows the kind of circuit shape that may be found in these very wide bandwidth active probes [5]. The probe tip contains no active components, and can be made extremely small for this reason. All it does contain is the parallel combination of 10 kΩ and 0.5 pF, which gives a low frequency attenuation of 46 db into the 50 Ω cable, when this is terminated with 50 Ω at the far end. At a frequency close to 30 MHz, which is $1/(2\pi \times 0.5 \text{ pF} \times 10 \text{ k}\Omega)$, the attenuation introduced by the probe tip network begins to fall at 6 db/octave.

It follows that the connector box needs to include an amplifier that, ideally, has a 50 Ω input impedance and a gain of 46 db at low frequency

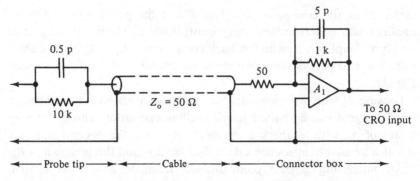

Fig. 5.9. *An active voltage probe for very wide bandwidth.*

which then falls at 6 db/octave beginning at the same critical frequency close to 30 MHz. Now this would call for the feedback components across A_1, in Fig. 5.9, to be 10 kΩ and 0.5 pF, identical to the values in the probe tip. Such an arrangement would give an overall gain of unity up to, and above, 30 MHz. However, a gain of 46 db over a bandwidth of 30 MHz is not an easy specification to meet, and when it is coupled with the requirement that the gain falls off at 6 db/octave from 30 MHz to around 1 GHz, it is virtually impossible, even with today's most advanced gallium arsenide microcircuit technology. For this reason the specification is relaxed to make the active probe shown in Fig. 5.9 have an overall attenuation of 20 db. It is, in fact, a ' × 10' active probe. Even with this relaxed specification, the amplifier, A_1 in Fig. 5.9, is a real challenge to the circuit designer. The design described by Rush, Escovitz and Berger [5] used four discrete microwave transistors in a thick film hybrid circuit [6].

5.10 Passive current probes

While the measurement of voltages, in a circuit under test, is most conveniently done with some kind of high impedance probe, the measurement of current calls for a very low impedance to be *inserted* in series with the current carrying conductor. For fairly high frequency work, the simplest way of doing this is to use a clip-on current transformer probe. These first became available around 1960 [7] and consist of a simple wound U-shaped core which has a sliding I-shaped core over the top. In this way the core may be opened, the lead carrying the current can be inserted, and the core then closed.

The design of such a passive current probe is best considered without the complication of needing to open and close the core. In fact, this kind of probe is of little use when currents in a printed circuit board are being

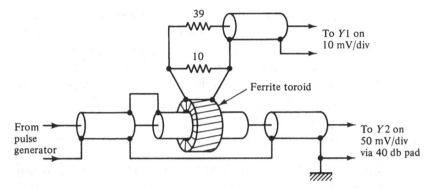

Fig. 5.10. *An experimental wide bandwidth current transformer. Full
details are given in the text. Note that the Y1 and Y2 input impedances
must be made 50 Ω and that the pulse generator must have 50 Ω output
impedance.*

studied. It follows that some circuits may need to have wide bandwidth
current transformers built into them during development. A good
example would be the measurement of the power dissipation in a
transistor as a function of time. It would be necessary to measure the
collector to emitter voltage with a high impedance, active or passive,
voltage probe, and also to measure the emitter current by including a wide
bandwidth current transformer around the emitter lead. A modern
digitising oscilloscope may have sufficient computing power to display
instantaneous power as a function of time by taking the product of these
two measurements. Otherwise, an analog multiplier could be used, or the
two waveforms simply displayed as Y1 and Y2, in the traditional way, and
the experimentalist would note the time at which, for example, the power
dissipation was at a maximum.

A valuable paper by Ritson and Wood [8] deals with the most
straightforward kind of wide bandwidth current transformer, and the
problem may be best considered by referring to Fig. 5.10. A ferrite toroid
is wound with a single layer winding of several turns, and the conductor
carrying the current to be measured passes through the centre of this
toroid. Between the winding and the current carrying conductor, some
kind of screen must be arranged, and in the experimental set-up shown in
Fig. 5.10, this is done by using a very short length of co-axial cable with
the outer conductor grounded at only one end. All the co-axial cables
shown in Fig. 5.10 are 50 Ω characteristic impedance.

The current transformation ratio of this kind of transformer is equal to
the number of secondary turns, N. The most convenient size of ferrite
toroid for the experiment shown in Fig. 5.10 is 9 mm [9]. This can carry

a uniform single layer winding of 50 turns, using 200 μm polyurethane coated wire, and then be a sliding fit on the outside of standard RG58C/U co-axial cable. Choosing a low value of secondary load resistor, R_L, not only reduces the insertion impedance of the transformer, which ideally is R_L/N^2, but also improves the bandwidth. Ritson and Wood [8] show that the low frequency cut-off frequency is approximately $R_L/2\pi L_s$, where L_s is the secondary inductance, and that the high frequency cut-off occurs at $1/2\pi C_s R_L$, where C_s is the stray capacitance across the secondary winding.

By choosing the 10 Ω and 39 Ω ($\approx$ 40 Ω) shown in Fig. 5.10, the 50 turn winding supplies only 1/500th of the primary current to the 50 Ω termination at the input to the $Y1$ amplifier on the oscilloscope. This is particularly convenient because the use of a 40 db attenuator pad, and the sensitivity ranges shown in Fig. 5.10, then allow a direct comparison of the input current to the transformer on $Y2$, and its reproduction of this current as a waveform on $Y1$. The pulse generator should be set for 5–10 V pulse amplitude.

The importance of the screen between primary and secondary may be checked easily with the set-up shown in Fig. 5.10 by temporarily disconnecting the connection to the outer conductor of the short length of co-axial cable that passes through the toroid. If the whole assembly is made in between two BNC panel sockets, it is possible to see the residual problems that remain, even when the screen is properly grounded, by reversing the direction of the assembly relative to the rest of the equipment. This changes the sign of the pulse signal observed on $Y1$, which is due to the magnetic coupling, but it does not change the sign of the spurious signal due to the unwanted capacitive coupling.

Such a passive current probe cannot be made to work up to very high frequencies. This is because of the poor permeability of the ferrite at higher frequencies, and also because of the length of the secondary winding itself. At very high frequencies the transformer must be designed as a distributed circuit. The secondary winding must be made in the form of a transmission line and terminated with the correct impedance. When this is done, it is possible to make wide bandwidth current transformers, for insertion into 50 Ω transmission line systems, which operate satisfactorily up to 1 GHz [10, 11].

5.11 Active current probes

One severe disadvantage of the simple current transformer, described in the previous section, is its inability to handle direct current and very low frequencies. This has been a problem in power engineering for a considerable time, and is solved by some kind of current comparator technique [12] in which the magnetic field in a toroidal core is nulled by a sensor and feedback arrangement. This causes the secondary ampere-turns to be exactly equal and opposite to the primary ampere-turns. Some active current probes of this kind can operate up to the high audio frequencies [13].

For really wide bandwidths, it is possible to combine a Hall sensor and the kind of wide bandwidth current transformer described in the previous section, all in one unit [14], and make the Hall sensor look after the d.c. and low frequency end of the spectrum. Such active current probes need very careful setting up and are sensitive to stray magnetic fields. Models are available which cover d.c. to 50 MHz [15].

5.12 Input circuits for power measurements

Voltage and current measurements are not at all easy at very high frequencies. For noise figure and gain measurements, the measurement of power, instead of voltage or current, may be a much more accurate approach.

This chapter closes with an example of one of the input circuits that can be of great use in power measurement: the directional coupler or directional bridge. This is a device that can be used to measure the incident or the reflected power at any input or output port of a network, device or system. Such a device also makes the measurement of input or output impedance, at high frequencies, particularly simple, and an example will be given in chapter 10. Incident and reflected power measurements are also needed when the magnitude of a network's S-parameters [16] are to be measured.

The function of the directional coupler, or directional bridge, may be seen at a glance from Fig. 5.11. The device has three ports. One is connected to a source of high frequency power, P_{in}, such as a signal generator or a swept frequency source. This source, of course, has an output impedance, usually 50 Ω, that matches the co-axial transmission line being used to interconnect the system under test.

The device under test, for example, a wide-band amplifier, is then connected to the port labelled 'load' in Fig. 5.11. A sensitive detector, with a 50 Ω input impedance, is connected to the third port. As shown in Fig. 5.11, any power which is reflected from the load, due to it not being a perfect match, is coupled back to the detector port where it can be measured.

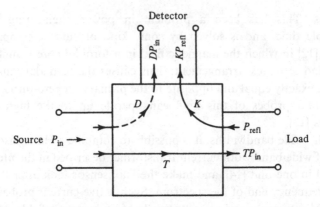

Fig. 5.11. *The directional coupler. The transmission loss, T, coupling factor, K, and the directivity, D, are discussed in the text.*

The directional coupler is a passive device so that T, K and D, shown in Fig. 5.11, must all be attenuations and these are usually expressed in decibels. Ideally, the transmission loss will be negligible, the directivity as high as possible, and the coupling factor only a few decibels. However, to get really high directivity, it is usually necessary to make the coupling factor, K, about 20 db. This is a nuisance when the directional coupler is being used to measure reflected power (the mode illustrated in Fig. 5.11) but, of course, 20 db is a sensible coupling factor when the same directional coupler is turned around and used to measure incident power.

Ideal directional couplers can only be made over a fairly limited bandwidth, and then only at quite high frequencies, because all must involve coupled waveguides or transmission lines at least one wavelength long. For really wide bandwidth and useful properties down to frequencies of only a few megahertz, directional couplers must make use of the kind of transformer which has been discussed above in section 5.9 and illustrated in Fig. 5.10. Spaulding [17] has published a very interesting design for such a directional coupler which will work over the range 1–1000 MHz, and the circuit shape of this coupler is shown in Fig. 5.12.

When the load port in Fig. 5.12 is terminated with 50 Ω, it is clear that no signal will appear at the detector port when power flows from source to load. This follows because power flow in this direction implies current flow from left to right during the positive half cycle of source and load voltage. The two transformers then produce equal and opposite voltages across the detector port which cancel one another out.

Power flow from load to source, however, implies current flow from right to left in Fig. 5.12 during the same positive half cycle of voltage. The signals from the two transformers now add across the detector port, and

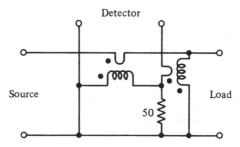

Fig. 5.12. *A wide bandwidth directional coupler using two transformers of the kind shown in Fig. 5.10. With turns ratio* 1:10, *the coupling factor, K, is* 20 *db.*

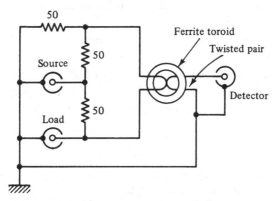

Fig. 5.13. *An experimental directional bridge for work on* 50 Ω *systems. For clarity, only one turn of the twisted pair around the ferrite toroid is shown.*

a 20 db fraction of this reverse power is available at the detector port when this is terminated correctly with 50 Ω.

The kind of wide bandwidth directional coupler which is shown in Fig. 5.12 can be made to have very small transmission loss and excellent directivity, but the construction of the transformers calls for very advanced techniques [17]. For many applications, a simpler device known as a 'directional bridge' [18] may be useful, and an example is shown in Fig. 5.13. This is the final experimental exercise for this chapter.

In Fig. 5.13 a classical bridge network is formed from three 50 Ω resistors and the load. Because one side of the load is grounded, one node of the bridge network becomes grounded, and the source can only be connected to the node which is opposite this grounded node. This leaves the other two nodes of the bridge free, and it is between these that the detector output can be taken. Because one side of the detector is grounded, the connection between the detector and bridge must be made *via* a balanced to unbalanced transformer of the kind discussed in chapter 3, sections 3.8 and 3.9. In this experimental directional bridge, the balun is again made using a 500 mm length of twisted pair, 0.4 mm polyurethane coated wire twisted to 5 mm pitch, wound around a 25 mm ferrite toroid [19].

Despite its name, the circuit shown in Fig. 5.13 does not have directional properties in the same way that the directional coupler shown in Fig. 5.12 has: source and load cannot be interchanged in Fig. 5.13 to obtain an incident power measurement. Source and detector ports in Fig. 5.13 may be interchanged, however. There are other ways of making these so-called directional bridges [20] so that they do have real directional properties, but these designs again call for wide bandwidth transformers with large turns ratios.

The experimental directional bridge shown in Fig. 5.13 should have a transmission loss of 6 db and a coupling factor of 6 db. The directivity of the bridge should be 60 db at 5 MHz, and fairly constant over the band 1–10 MHz. It is a very useful device for checking impedance in circuits and devices which are intended for use in 50 Ω coaxial systems.

Notes

1 Horowitz, P., and Hill, W., *The Art of Electronics*, Cambridge University Press, Cambridge, second edition, 1989, pp. 1048–9.
2 McGovern, P. A., *IEEE Trans. Instr. Meas.*, **IM-26**, 46–52, 1977.
3 These problems have been discussed by V. Bunze in *Probing in Perspective*, Hewlett–Packard Application Note No. 152, 1972, and by J. M. Williams in an Appendix to Linear Technology Corp. Application Note No. 13, April 1985.
4 This circuit is essentially the one used in the Hewlett–Packard 1120A 500 MHz active probe, which was introduced in 1970 and was still available in 1985.
5 Rush, K., Escovitz, W. H., and Berger, A. S., *Hewlett–Packard J.*, **37**, No. 4, 11–19, April 1986.
6 Toeppen, D. E., *Hewlett–Packard J.*, **37**, No. 4, 33–6, April 1986.
7 Forge, C. O., *Hewlett–Packard J.*, **11**, Nos. 11 and 12, July and August 1960.
8 Ritson, F. J. U., and Wood, J. *Electronic Engineering*, **36**, 483–5, 1964.
9 The Mullard FX3240 high permeability 9 mm toroid is suitable for this. So are the Siemens 10 mm toroids in N30 and T38 Siferrit material.
10 O'Dell. T. H., *Electr. Lett.*, **5**, 369–70, 1969.
11 Anderson, J. M., *Rev. Sci. Instr.*, **42**, 915–26, 1971.
12 Moore, W. J. M., and Miljanic, P. N., *The Current Comparator*, Peter Peregrinus Ltd., London, 1988.
13 Etler, J. P., and Soffer, J. S., *IEE Conf. Publ.* No. 264, 164–8, 1986.
14 Hongel, C., *Tektronix Service Scope*, Oct. 1967, p. 10.
15 For example, the Tektronix A6302 current probe.
16 Carson, R. S., *High-Frequency Amplifiers*, John Wiley Inc., New York, second edition, 1982.

17 Spaulding, W. M., *Hewlett–Packard J.*, **35**, No. 11, 17–20, Nov. 1984.
18 Ichino, T., Ohkawara, H., and Sugihara, N., *Hewlett–Packard J.*, **31**, No. 1, 22–32, Jan. 1980.
19 Mullard FX3312 Ferroxcube toroids are a good choice. So are the Siemens R25 toroids in type N30 Siferrit.
20 See the *Radio Amateur's Handbook*, ARRL, Newington, Conn., USA, 1981, pp. 16.11 and 16.31.

6
Wide-band amplifier circuits

6.1 Introduction

Wide-band amplifier circuit design is one of the most difficult and varied areas in the whole subject of electronic circuit design. There are the wide-band intermediate frequency amplifiers associated with radio, television and radar receivers, there are the wide-band pulse amplifiers of radar and nuclear instrumentation systems, there are the direct-coupled wide-band amplifiers found in oscilloscopes, and many other electronic systems, and there are the wide-band repeater amplifiers found in cable communication systems.

A number of excellent texts consider many of the areas listed above. Maclean's book [1] deals mainly with the repeater kind of amplifier, and is particularly strong on the computer aided optimisation of a design for the best noise performance and low sensitivity to component tolerance. Kovács' book [2] covers a much wider field and attempts to answer the question which will be central to this chapter: how is the first step in design, the choice of the *circuit shape*, the initial circuit idea, to be made? This is, perhaps, in contrast to Carson's well-known book [3], which contains no practical circuits but is excellent on theory.

This chapter will concentrate on one small area in the field of wide-band amplifier circuit design: wide-band direct-coupled amplifiers. These are needed for the deflection amplifiers of the conventional analog oscilloscope and, of course, for the digitising oscilloscope when this kind of instrument is to be designed for use with very low level signals. In both cases, the input stages of the deflection amplifier call for high input impedance, wide bandwidth and well-defined gain.

The output stage of an analog oscilloscope deflection amplifier is also of particular interest, when the CRT is of the conventional kind, because

90

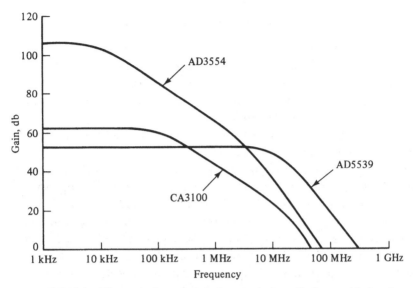

Fig. 6.1. *The open-loop gain characteristics of three wide-band operational amplifiers: the hybrid AD*3554, *the monolithic BiMOS CA*3100, *and the pure npn monolithic AD*5539.

the load will be a pure capacitance which must be driven at high frequency and at quite high voltage. The output stage problem is considered at the end of this chapter.

6.2 Direct-coupled amplifiers

A direct-coupled amplifier, or d-c amplifier, is one which has a constant gain from zero frequency up to some high frequency at which the gain begins to fall. This is illustrated in Fig. 6.1 where the open loop gain characteristics of three well-known wide-band operational amplifiers are shown [4]. The contrast between these amplifiers and the more common kind of internally compensated operational amplifier, which has a constant gain only up to about 10 Hz, should be noted.

Wide-band d-c amplifiers are a relatively recent kind of circuit in electronics. The first analog oscilloscopes used amplifiers with *RC* coupling, and could therefore only display time varying signals. There was no possibility of using the oscilloscope to display relative d.c. levels in a circuit. These early oscilloscopes used several stages of amplification, without any negative feedback, and had inductive loads for the active devices in order to compensate for the fall off in gain at high frequencies [5]. The first video amplifiers of early television followed the same lines.

6.3 Wide-band d-c amplifiers for oscilloscopes

The simplest analog oscilloscope must have a bandwidth of well over 10 MHz if it is to be of any use to the electronic engineer of today. The sensitivity of the instrument should be such that signals of a few millivolts are sufficient to give a vertical deflection of about one screen diameter. This will call for an overall gain of at least 80 db.

It is thus clear that the operational amplifiers which have the kind of open-loop gain characteristics shown in Fig. 6.1 are not going to be the building blocks of a modern analog oscilloscope amplifier. Even the AD5539 would only be useful at the front end of an oscilloscope with a modest bandwidth of about 20 MHz, and then something would have to be done about its high input bias current. A very important feature which any oscilloscope must have is that the apparent d.c. level displayed is quite independent of the signal source impedance. When the input socket to the oscilloscope is short circuited, open circuited, or connected to some finite impedance, there must be no observable shift in the display. This clearly rules out any kind of amplifier which has a significant input bias current.

6.4 The problem of the input circuit

The input attenuator circuits for a typical analog oscilloscope were discussed in chapter 5, section 5.6. These were designed to work into an amplifier having a 1 MΩ input impedance. On the most sensitive range of the instrument, the input signal is taken directly to this 1 MΩ input impedance circuit.

As stated in the previous section, the input circuit should have negligible input bias current. It must also be protected in some way from the possibility that the user will connect the input of the oscilloscope to some very high voltage.

Both these essential requirements are satisfied by means of the circuit shown in Fig. 6.2. In this typical analog oscilloscope circuit, a dual JFET, a 2N5911, is used. This kind of device has a very small input bias current, typically 100 pA at 25 °C. This will produce a change in apparent input voltage of less than 100 μV when the input socket is short circuited. Unfortunately the input bias current of a JFET increases very rapidly with temperature and, from this point of view, a MOSFET would be better. However, the MOSFET suffers from greater flicker noise, a point which will be discussed in chapter 10, and it is more difficult to get a matched pair of devices.

The 2N5911 is protected by means of the two 1N914 diodes in Fig. 6.2.

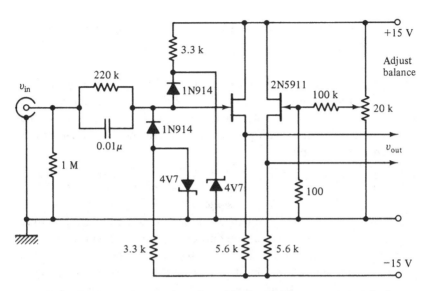

Fig. 6.2. A typical input stage for an analog oscilloscope using a dual JFET which is protected against excessive input voltages.

If the oscilloscope input, on its most sensitive range, which would be typically a few millivolts per division, were connected to a voltage in excess of ± 4.7 V, one or the other of the 1N914 diodes would begin to conduct and limit the gate to source voltage of the 2N5911 to a value well below breakdown. In fact some ± 600 V could be applied to the input before the protection circuit would be rendered ineffective, because of the 220 kΩ resistor which has been put in series with the input lead. This 220 kΩ resistor is shunted with a 0.01 μF capacitor to eliminate the attenuation that it would present to high frequency input signals because of the capacitance at the gate of the 2N5911. This capacitance can total about 10 pF: the input capacitance of the 2N5911 plus the capacitance of the two reverse biassed 1N914 diodes. Typically, the total input capacitance of a simple analog oscilloscope will total about 20 pF because of additional wiring and because of the problems discussed in chapter 5, section 5.6.

6.5 Feedback

The gain of the input stage shown in Fig. 6.2, $\Delta v_{\text{out}}/\Delta v_{\text{in}}$, is very close to unity. If R_{s} is the 5.6 kΩ resistor connected to the source of the left-hand side of the 2N5911, it is clear that

$$g_{\text{m}}(\Delta v_{\text{in}} - \Delta v_{\text{out}}) = \Delta v_{\text{out}}/R_{\text{s}} \qquad (6.1)$$

where g_{m} is the forward transconductance of the 2N5911.

Equation (6.1) leads to the well-known result for the gain of a source follower circuit,

$$\Delta v_{\text{out}}/\Delta v_{\text{in}} = g_{\text{m}} R_{\text{s}}/(1 + g_{\text{m}} R_{\text{s}}) \qquad (6.2).$$

This is, of course, evidence that the source follower is really an amplifier with an open-loop gain of $g_{\text{m}} R_{\text{s}}$ which has 100 % negative feedback. The 100 % negative feedback arises because of the *circuit shape*: the output terminal and the negative going input terminal are one and the same thing, the source of the JFET.

Feedback of this very local kind is a feature of the circuits which will be considered in the next few sections as a typical analog oscilloscope vertical deflection amplifier is dealt with stage by stage. Fig. 6.2 shows another feature which will be common to all these circuits: balance. The right-hand side of Fig. 6.2 provides a second output terminal which should be at the same mean voltage level as the left-hand output terminal, regardless of changes in temperature. The stages which follow should then take this balanced output signal and amplify it up to the level required for the vertical deflection plates, which present a balanced load.

Local feedback, that is negative feedback across each stage of the vertical deflection amplifier, is essential if the full gain-bandwidth of the active devices is to be reflected in the overall bandwidth of the amplifier. This is clear if Fig. 6.1 is considered again. The AD3554 and AD5539 operational amplifiers have gain-bandwidth products well over 1 GHz, as would be expected from a bipolar monolithic or hybrid technology, but, because these are both multistage amplifiers, only a modest amount of feedback may be applied overall, otherwise the resulting feedback amplifier will be unstable. This is particularly evident for the AD3554: despite its 1.7 GHz gain-bandwidth product, inferred from an extrapolation of the 6 db/octave part of its open-loop gain characteristic, it cannot be used with a feedback fraction of more than − 60 db, giving a gain of 60 db but a bandwidth of only about 1 MHz.

6.6 Local series and shunt feedback

Very wide bandwidth and well-defined gain can be obtained from circuits which are built up from the two elementary circuit shapes shown in Figs. 6.3(*a*) and (*b*). These are often called 'series feedback' and 'shunt feedback' circuits, respectively [6].

In the series feedback circuit the transistor is being used as a transconductance amplifier, which means that its output current is considered to depend upon its base to emitter voltage. This voltage is formed by subtracting a voltage proportional to the output current from

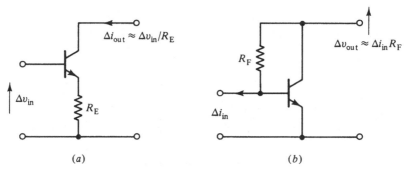

Fig. 6.3. *The series feedback circuit, (a), and the shunt feedback circuit, (b).*

the true input voltage, v_{in}. In other words, this is an example of negative feedback of a *voltage* proportional to *output current*. It results, as shown in Fig. 6.3(a), in a circuit which converts a change in input voltage into a well-defined change in output current: $\Delta i_{out} \approx \Delta v_{in}/R_E$. An accurate expression will be given below, when the final circuit shape is discussed.

In the shunt feedback circuit, Fig. 6.3(b), the transistor is being used as a current amplifier, which means that its output current is considered to depend upon its base current. This base current is formed by subtracting a current proportional to output voltage from the true input current, i_{in}. In other words, this is an example of negative feedback of a *current* proportional to *output voltage*. It results, as shown in Fig. 6.3(b), in a circuit which converts a change in input current into a well-defined change in output voltage: $\Delta v_{out} \approx \Delta i_{in} R_F$. Again, an accurate expression is given below.

6.7 Combining the shunt and series feedback circuits

The simple way in which Figs 6.3(a) and (b) have been drawn suggests, almost at once, that these two circuits can be combined to form a new circuit shape for a voltage amplifier with a gain R_F/R_E. This idea may have been first put forward in an important paper by Cherry and Hooper [7] and is shown in Fig. 6.4, but Fig. 6.4 is a long way from being any kind of useful circuit. In the first place, a balanced amplifier is needed for the analog oscilloscope deflection amplifier, for the reasons given in section 6.5. Secondly, a practical circuit based upon Fig. 6.4 would introduce a considerable level shift because the output terminal must be positive, with respect to the input terminal, by $2V_{CE}$, where V_{CE} will be at least 5 V if good high frequency performance is to be obtained. This level shift may not matter in an analog oscilloscope deflection amplifier because the CRT

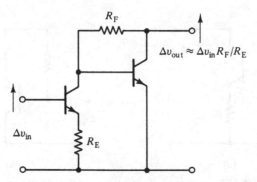

Fig. 6.4. *An academic series/shunt feedback pair.*

deflection plates could both be at quite a high d.c. level. It is unlikely, however, that the designer would consider this way of accommodating a large level shift a good idea, because it would involve quite complicated power supply arrangements for the amplifier circuits.

6.8 An experimental circuit

Fig. 6.5 shows the first experimental circuit for this chapter and illustrates one of the many possible ways of providing balanced amplification with negligible level shift by combining the series feedback and shunt feedback circuits.

In Fig. 6.5, series feedback is applied to the long tailed pair Q_5 and Q_4 by means of two equal resistors, R_3 and R_4. The current output from this long tailed pair drives a second long tailed pair, Q_2 and Q_1, which has shunt feedback applied by means of two equal resistors, R_{11} and R_{15}. The circuit is built around a CA3096A transistor array, and the numbering of the devices corresponds to the numbering on the data sheet.

By using pnp transistors for the first stage, Q_5 and Q_4, and then npn transistors for the second stage, Q_2 and Q_1, a negligible level shift is involved in going from input to output when the component values are chosen correctly. There are, of course, a number of ways of doing this when an amplifier involves several stages, instead of simply two stages like the circuit shown in Fig. 6.5. The first stage may be npn, the second stage pnp: Fig. 6.5 uses the alternative simply because the pnp devices in the CA3096A are designed to work at a lower current than the npn devices, so that it is sensible to use pnp at the input end and npn at the output. Another possible mix of device polarity is to use npn, or pnp, for the first two stages, and then restore the level by using the opposite polarity in the third and fourth stages, and so on.

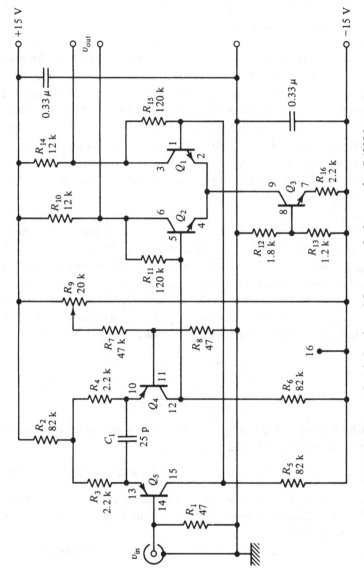

Fig. 6.5. An experimental wide-band amplifier circuit built around a CA3096 transistor array.

This idea of mixing device polarity to give zero d.c. level shift between the input and output terminals of a d-c amplifier is very old. A circuit published in 1955 by Slaughter [8] uses the idea, with transistors of the very earliest kind. The wide adoption of this idea had to wait for really good dual high frequency devices, of both pnp and npn polarity, to become available. A monolithic process which can provide good transistors of both polarities still presents problems. This is why many analog oscilloscope vertical amplifiers are made with discrete components, albeit using the most advanced surface mounting techniques with components on both sides of the circuit board [9]. The operational amplifiers which are the subject of Fig. 6.1 solve the level shifting problem as follows: the AD3554 is a hybrid circuit using npn and pnp devices; the CA3100 uses npn devices and p-channel MOSTs; the AD5539 has only npn devices and makes use of several base-emitter voltage drops to give the required level shift.

The resistors R_5 and R_6 play an important role in the circuit shown in Fig. 6.5 because they must carry the mean collector currents of Q_5 and Q_4 as well as the mean currents flowing in the shunt feedback resistors, R_{11} and R_{15}. In most designs these two currents are about the same order of magnitude. The currents in R_5 and R_6 are almost constant because changes in the collector currents of Q_5 and Q_4 should be transferred directly into the shunt feedback stage, Q_2 and Q_1. Writing $R_3 = R_4 = R_E$, these changes in collector current for Q_5 and Q_4 may be written,

$$\Delta i_{c5} = -g_{m5}(\Delta v_{in}/2)/(1+g_{m5} R_E) \tag{6.3}$$

and

$$\Delta i_{c4} = +g_{m4}(\Delta v_{in}/2)/(1+g_{m4} R_E) \tag{6.4}$$

Writing $g_{m5} = g_{m4} = g_m$, it follows that a fairly accurate expression for the input current from Q_2 and Q_1 would be,

$$\Delta i_{c4} - \Delta i_{c5} = \Delta v_{in}/(g_m^{-1} + R_E) \tag{6.5}$$

and this will only be equal to $\Delta v_{in}/R_E$ when the term g_m^{-1} is small compared to R_E. For this reason it is important to keep the full expression, equation (6.5), in the analysis.

Passing on to the second stage, it is easy to show, using equation (6.5), that

$$\Delta v_{out}(1/R_F + 1/R_L h_{fe} + 1/R_F h_{fe}) = \Delta v_{in}/(g_m^{-1} + R_E) \tag{6.6}$$

where $R_F = R_{11} = R_{15}$, $R_L = R_{10} = R_{14}$ and $h_{fe} = h_{fe1} = h_{fe2}$.

The term $1/R_F h_{fe}$ can be neglected, but if R_L is made much lower than R_F the feedback resistor should be thought of as being shunted by a resistor $R_L h_{fe}$. A fairly accurate expression for the overall low frequency gain of the circuit shown in Fig. 6.5 is thus

$$\Delta v_{out}/\Delta v_{in} = R'_F/R'_E \qquad (6.7)$$

where R'_F is the parallel combination of R_F and $R_L h_{fe}$, and R'_E is the series combination of R_E and g_m^{-1}.

6.9 Component values

All the component values are marked on Fig. 6.5. A single ended input is provided, acting as a termination for the pulse generator that will be used as a signal source. This means that the 'adjust balance' potentiometer, shown first in Fig. 6.2, is now transferred to the experimental circuit as R_9.

R_2 is chosen to make the collector currents in Q_5 and Q_4 about 100 μA each, because this is the optimum for the pnp devices in the CA3096 array. This makes g_m^{-1}, in equation (6.7), about 300 Ω so that R_E is chosen to be well above this, but not so great as to allow g_m^{-1} to be neglected.

Q_3 is being used as the current sink for Q_2 and Q_1 and, as these npn devices in the CA3096A have optimum h_{FE} at a collector current of about 1 mA, R_{12}, R_{13} and R_{16} are chosen to make Q_3 sink just over 2 mA.

The shunt feedback resistors are chosen to give an overall low frequency gain of about 50, giving $R_{11} = R_{15} = 120$ kΩ. The V_{CE} of Q_5, Q_4, Q_2 and Q_1 is then chosen to be 5 V in magnitude and this enables R_5 and R_6 to be determined because the d.c. current in R_{11} and R_{15} is then known. The values of R_{10} and R_{11} then follow, and as these can be made 12 kΩ it is clear that $h_{fe} R_L$ is much greater than R_F, in equation (6.7), at least at low frequency, and the low frequency gain of the experimental circuit should be given by

$$G_0 = R_F/(g_m^{-1} + R_E) \qquad (6.8)$$

which comes out to be 48.

6.10 Experimental measurements

Initially, measurements should be made without C_1 connected. Check the d.c. levels in the circuit and then connect an oscilloscope to the output points, pins 3 and 6 of the CA3096A, using two $\times 10$ high impedance probes. The oscilloscope sensitivity should be at least 100 mV/div and bandwidth at least 10 MHz.

Apply a 10 μs positive going pulse of a few millivolts and check that the low frequency gain is close to the value of 48 predicted by equation (6.8).

The full gain is, of course, given by the true differential output voltage, observed between pins 3 and 6 on the CA3096A, divided by the input pulse amplitude. A positive pulse should be seen on pin 3, and an equal, but opposite polarity, pulse on pin 6. The experimentalist should resist the temptation at this stage of using the oscilloscope in the 'channel A plus inverted channel B', or differential, mode, because some interesting things may be seen which need explaining. For example, when the output pulses are observed separately, removing one of the high impedance probes causes the pulse which is left under observation to have a much longer rise time. In other words, adding capacitance to pin 3 slows down the response of pin 3, as would be expected, but appears to speed up the response on pin 6, the explanation of which may need some thought. The circuit is quite symmetrical in this respect: a few picofarads added between pin 6 and ground, when both pin 3 and pin 6 are under observation, will be seen to slow the response at pin 6 while speeding the response at pin 3.

These remarks lead to a discussion of the high frequency performance of the experimental circuit shown in Fig. 6.5. It is obvious that the shunt feedback stage, Q_2 and Q_1, will be very sensitive to any capacitance which may be in parallel with the feedback resistors, R_{11} and R_{15}. The R_F, which occurs in equation (6.6), should be replaced by Z_F, to represent the parallel combination of R_F and C_{cb}, where C_{cb} is the collector junction capacitance of Q_1 and Q_2, plus any stray capacitance due to the circuit hardware. As it stands, this output stage should have a rise time of the order of $C_{cb} R_F$, which could be as long as 200 ns.

However, it will be observed that the experimental circuit shown in Fig. 6.5 is not very sensitive to capacitive loading across its true output: that is from pin 3 to pin 6. The reason for this is the quite low differential output impedance. The experimentalist should obtain an expression for this impedance, which should simplify to $R_{out} = 2R_F/h_{fe}$ under the same assumptions which lead to equation (6.8), and make a measurement of R_{out} by adding a resistive load from pin 3 to pin 6.

The addition of C_1, shown in Fig. 6.5, to the experimental circuit makes a considerable improvement to the speed of response because this should compensate for the $C_{cb} R_F$ time constant of the output stage by means of the $2C_1 R_E$ time constant introduced at the input stage. The circuit is then left with two other very serious high frequency limitations which have not been mentioned so far. The first is the frequency dependence of h_{fe} for Q_2 and Q_1 and brings the argument back to equation (6.7). The h_{fe} of the npn devices in the CA3096A begins to fall off above 4 MHz, and it will no longer be possible to assume that $h_{fe} R_L$ is much greater than R_F. However, with a low frequency h_{fe} of 400 and R_L at 12 k it will be at frequencies well above 10 MHz before $h_{fe} R_L$ falls to equal R_F. The gain of the circuit would then begin to fall very rapidly, as does h_{fe}.

The second reason for the limited bandwidth of the circuit shown in Fig. 6.5, and the most important, is the poor high frequency performance of the pnp transistors in the CA3096A: Q_5 and Q_4. These lateral devices have a gain-bandwidth product of only 6.8 MHz and, in fact, this will be found to be about the bandwidth of the experimental circuit when C_1 is in place.

By using the CA3096A, it has been possible to build an experimental circuit of the same circuit shape as a real analog oscilloscope deflection amplifier. The high frequency performance limits, however, have been brought down into a lower frequency range where the experimentalist may make measurements with quite simple equipment.

A complete understanding of the high frequency performance of the experimental circuit shown in Fig. 6.5 calls for careful modelling of the active devices and detailed calculation. For a first consideration of this problem, chapter 7 of the book by Gray and Meyer [6] is helpful. To take the problem further, an excellent paper by Faulkner [10] will be very useful.

6.11 Output stage circuits

The output stage of an analog oscilloscope deflection amplifier calls for some special consideration because of the nature of the load which is put upon it. When the CRT is a conventional one, with electrostatic deflection, this load is a pure capacitance which could be as much as 20 pF with the cables that will be involved [11], and full vertical deflection could call for differential output voltages of 50 V amplitude. For a quite modest oscilloscope rise time of 10 ns, this then calls for a maximum output current, $i = C \, \mathrm{d}v/\mathrm{d}t$, of at least 100 mA from the deflection amplifier.

Very wide bandwidth analog oscilloscopes, that is instruments with bandwidths above 250 MHz, do not have the same problems because their CRTs will have a distributed vertical deflection system [12] which presents a constant and real impedance load for the wide-band deflection amplifier. The horizontal deflection amplifier will be of the type discussed here, however.

Consider Fig. 6.6 as a first step towards a solution of the problem of finding a good circuit shape for the output stage of an analog oscilloscope deflection amplifier. In Fig. 6.6, v_{in} should be thought of as the differential output from several stages of amplification of the kind discussed so far in

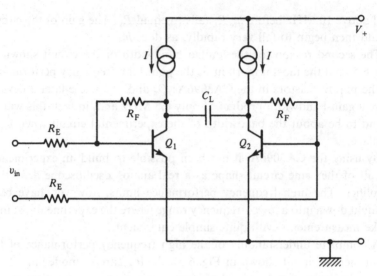

Fig. 6.6. *A first step towards a good circuit shape for the output stage of an oscilloscope deflection amplifier. The CRT deflection plates are represented by the capacitor, C_L.*

this chapter and illustrated by the experimental circuit shown in Fig. 6.5. The proposed output stage shown in Fig. 6.6 has a well-defined gain, R_F/R_E, and v_{in} would have a small negative common mode level, $-V_+ R_E/2R_F$, in order to bring both ends of the load capacitor, C_L, which represents the CRT deflection plates, to a voltage level $V_+/2$. It is not, of course, necessary to level shift the voltage which is actually at both deflection plates so that this voltage has zero mean level.

Now as Fig. 6.6 stands, the constant current sources which are used as collector loads for Q_1 and Q_2 will have to have a magnitude slightly greater than the maximum current demanded by the deflection plates. This was calculated to be 100 mA at the beginning of this section. Similarly, V_+ will have to be slightly greater than the maximum deflection voltage mentioned above, and this was 50 V. The argument so far is leading towards an output stage that dissipates 10 W when it is under quiescent conditions and when its true load, C_L, is taking no current at all. This is clearly a point where the circuit designer stops to consider what is really called for in this circuit.

Two points should be taken into consideration. The first is that the load, C_L, calls for negligible current for most of the time because any analog oscilloscope is, for most of the time, displaying signals which change with time very much more slowly than the maximum rate of change that the instrument is able to display. The second point is that

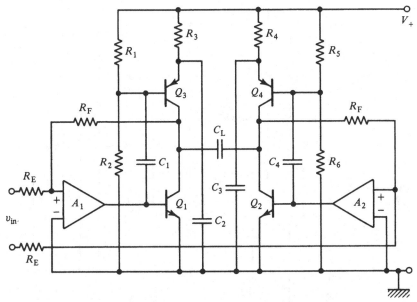

*Fig. 6.7. A second step towards a good circuit shape for the output stage
of an analog oscilloscope deflection amplifier.*

when rapid deflections are called for, the high transient current demanded
by C_L may be drawn from a capacitor somewhere in the circuit. Provided
this capacitor is much greater than C_L, it will always be able to supply the
transient current because rapidly repeated transients could only cause
problems when these occur so frequently that the input signal frequency
is beyond the bandwidth of the instrument.

These two considerations suggest a development from the circuit shape
shown in Fig. 6.6 to the new circuit shown in Fig. 6.7. The details of the
current sources for Q_1 and Q_2 are now shown. These current sources, Q_3
and Q_4, both supply d.c. at a level determined by the choice of R_3 and R_4,
and the voltage level set at the bases of Q_3 and Q_4 by means of the
potential dividers, $R_1/(R_1 + R_2)$ and $R_5/(R_5 + R_6)$.

While Q_3 and Q_4 supply a d.c. of only a few milliamps, the large
transient currents would be available because of the way in which the
capacitors C_2 and C_3 have been arranged. This, along with the fact that
the base of Q_3 is now connected, at high frequencies, to the base of Q_1
through C_1 (and, similarly, Q_4 is connected to Q_2 through C_4), means that
Q_3 can supply a large current, out of C_2, when Q_2 draws this same large
current from the right-hand side of C_L. Similarly, Q_4 can supply the large
current, out of C_3, which is called for by Q_1 when current is to be drawn
from the left-hand side of C_L.

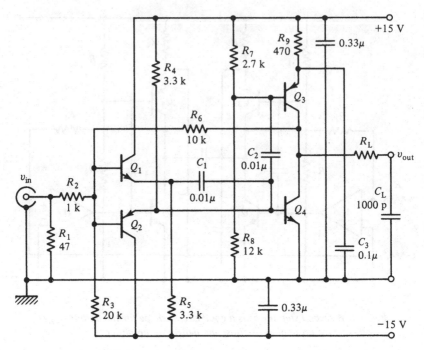

Fig. 6.8. *An experimental output circuit.* Q_1 *and* Q_4 *are* 2N2222A *and* Q_2 *and* Q_4 *are* 2N2907A. *For the first measurements omit* C_1 *and* C_2 *and make* R_L *zero.*

6.12 An experimental output circuit

Fig. 6.8 shows an experimental version of the circuit shown in Fig. 6.7. Only one side of the output stage need be built, because all the interesting behaviour can be observed when one end of the load, C_L, is grounded and the other end is driven, either positive or negative going, from a mean level of $V_+/2$.

To begin with, R_L should be made zero and C_L made 1000 pF. The operational amplifier, A_1 in Fig. 6.7, has been replaced in Fig. 6.8 by simple emitter followers, Q_1 and Q_2. These emitter followers provide a symmetrical low impedance drive to the output transistors, Q_3 and Q_4, and also, by making $R_3 = 2R_6$, set the d.c. level at the output terminal to $V_+/2$. The overall voltage gain is set by making $R_6/R_2 = 10$.

Two very valuable circuit shapes can be seen in Fig. 6.8. The first is the output stage itself, consisting of two complementary transistors with their *collectors* connected together to form the output terminal. The second is the connection of the two emitter followers, Q_1 and Q_2 in Fig. 6.8. These complementary devices have their *bases* connected together to form an input terminal (equivalent to the positive input of A_1 in Fig. 6.7). Unlike the well-known complementary emitter follower, however, the emitters of Q_1 and Q_2 are taken to quite separate resistors, R_4 and R_5, so that, under quiescent conditions, Q_1 and Q_2 both have the same magnitude of collector current.

Because Q_1 and Q_2 are npn and pnp respectively, their base currents are of opposite sign and this means that it is possible, when their current gains are about equal, for the quiescent current in R_2 to be very small. A further advantage of this circuit shape is that the two outputs have a level shift of about ± 0.7 V with respect to the common input. In Fig. 6.8 this is used to make the input terminal have a very small offset voltage: the level shift up to 0.7 V provided by Q_2 is just what is needed to direct-couple into Q_4 from an input terminal close to zero mean level.

The high frequency drive from Q_1 and Q_2 into Q_3 and Q_4 is though the capacitors, C_1 and C_2. It was pointed out above, in section 6.11 when discussing Fig. 6.7, that the quiescent current in the output transistors need only be a few milliamps. In the experimental circuit R_7, R_8 and R_9 have been chosen to make this quiescent current about 4 mA.

6.13 Measurements on the experimental output circuit

The power supply voltage used in the experimental circuit is only $+15$ V, in contrast to at least 100 V which would be found in any real analog oscilloscope deflection amplifier output circuit. This is done for experimental convenience. It is also convenient to make the circuit at least one order of magnitude slower than any real circuit so that measurements can be made with an oscilloscope of modest bandwidth.

This means that C_L should be at least two orders of magnitude greater than the 10 pF typical of a real circuit: one order for the lower supply voltage and one order for the lower circuit speed. The transient current in the experimental circuit will then correspond to that in the real circuit.

In view of this, C_L is made 1000 pF. To begin with C_1 and C_2 are omitted. This makes it possible to check the very slow transient performance of the experimental circuit when it is operating along lines analogous to Fig. 6.6 with constant current sources of only 4 mA.

The input signal needed is a pulse, 10 μs duration with an amplitude between ± 100 mV and ± 1 V. As the circuit has a gain of 10, it is possible to observe the maximum output swing, positive or negative, with an input pulse of less than ± 1 V amplitude. The slew rate at the output will be seen to be only a few volts per microsecond at this stage. The form of the response will be simply first order: an exponential rise or fall from $V_+/2$.

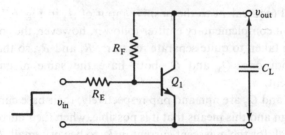

Fig. 6.9. *The essential features of the circuit shown in Fig.* 6.8 *when there is a positive going input step and* $R_L = 0$.

Now when C_1 and C_2 are added to the experimental circuit a dramatic improvement in speed naturally results. The rise and fall times at the output, across the 1000 pF load capacitor, C_L, will be only a few tens of nanoseconds. This means that currents of hundreds of milliamps are now available.

The form of the transient response, in this new fast circuit, is not at all satisfactory, however. The response to a step input will now be a damped oscillation: typically some 30% overshoot followed by several cycles of oscillation at a frequency around 10 MHz. It is important to understand why this is so, and then see how the circuit should be changed to correct for the poor transient performance.

6.14 Transient analysis

The reason for the underdamped transient response of the experimental circuit, Fig. 6.8, and the steps that should be taken to correct this, may be seen if the simple circuit of Fig. 6.9 is considered.

In Fig. 6.9, Q_1 should be thought of as representing both Q_1 and Q_4 in Fig. 6.8, because these are the two devices which are active when a positive input step, as shown in Fig. 6.9, is applied.

To analyse Fig. 6.9, the charge control equations [13]

$$i_b = q_b/T_b + dq_b/dt \qquad (6.9)$$

and

$$i_c = q_b/T_t \qquad (6.10)$$

are all that are required. These equations relate the instantaneous base current, i_b, and collector current, i_c, to the charge, q_b, in the base region of the device when it is working in the active state. The two characteristic time constants are, approximately, the minority carrier lifetime in the base, T_b, and the base transit time, T_t. In the steady state, dividing equation (6.10) by equation (6.9) leads to the relationship

$$T_b/T_t = h_{FE} \qquad (6.11)$$

for the device current gain at zero frequency.

Substituting equation (6.10) into equation (6.9), and using equation (6.11) leads to

$$T_t \, di_c/dt + i_c/h_{FE} = i_b \qquad (6.12)$$

as the differential equation governing the transistor.

Now in the simple circuit of Fig. 6.9, the current in R_F may be neglected in comparison to the collector current. It follows that

$$i_c = -C_L \, dv_{out}/dt. \qquad (6.13)$$

The base current of Q_1, in Fig. 6.9, is simply

$$i_b = v_{in}/R_E + v_{out}/R_F. \qquad (6.14)$$

Substituting equation (6.13) and (6.14) into equation (6.12) then leads to the second order differential equation which governs the entire circuit:

$$(C_L R_F T_t) d^2 v_{out}/dt^2 + (C_L R_F/h_{FE}) dv_{out}/dt + v_{out} = -(R_F/R_E) v_{in}. \qquad (6.15)$$

Equation (6.15) shows that the steady state output, $v_{out} = -(R_F/R_E)v_{in}$, will be approached by means of a quite lightly damped sinusoidal response. The damping time constant in equation (6.15) is $C_L R_F/h_{FE}$ and is small, because h_{FE} is the product of the current gains of both Q_1 and Q_4 in Fig. 6.8.

There is far more to this analysis than this simple conclusion, however. It is now possible to see what must be added to Fig. 6.9, and thus to Fig. 6.8, to increase the damping. The reason for the light damping is that C_L holds down the rise of the output voltage when the input step arrives, thus allowing the base current to have the very high initial value of v_{in}/R_E. What is needed is some modification to the circuit so that some feedback takes control right from the start of the transient, and there are two ways to do this.

The first thought the designer might have would be to add a capacitor in parallel with R_F. This capacitor, C_F, would begin to feed current, $C_F \, dv_{out}/dt$, back to the base of Q_1 immediately the voltage across C_L began to change.

A second approach for the designer would be to recall that this is a deflection amplifier output circuit, and that it is the voltage across C_L

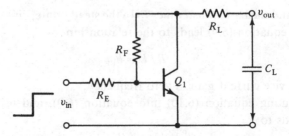

Fig. 6.10. *Adding resistor R_L increases the circuit damping.*

which is the output voltage of interest. Adding resistance, R_L, in series with C_L, as shown in Fig. 6.10, allows the collector voltage to change by an amount $i_c R_L$ immediately the collector current begins to flow. This change in collector voltage immediately produces a change in the current fed back through R_F. The equation governing Fig. 6.10 would then be

$$(C_L R_F T_t)\,d^2 v_{out}/dt^2 + C_L R_F(h_{FE}^{-1} + R_L/R_F)\,dv_{out}/dt + v_{out} = -(R_F/R_E)v_{in}$$
(6.16)

in contrast to equation (6.15). The damping has been increased.

Of the two methods, adding R_L is a better solution because it increases the damping without also reducing the natural resonant frequency of the circuit, which remains at

$$\omega_n = (C_L R_F T_t)^{-\frac{1}{2}}.$$
(6.17)

Naturally, this very approximate model may only be used as a rough guide to the value of R_L that will give optimum transient response. What is important about the model is that it should lead the designer to see how the original circuit should be modified in order to correct for the unacceptable performance that it originally had.

6.15 Further experiments and conclusions

Equation (6.16) suggests that R_L/R_F should be made about 10^{-3} for a good transient response to be obtained from the experimental circuit of Fig. 6.8. This will be found when R_F is taken at the value of 10 kΩ, given for R_6 in Fig. 6.8, C_L is 1000 pF and T_t is taken at 0.5 ns, which is a sensible value for the kind of devices being used.

From this rough calculation, values of R_L should now be added to the experimental circuit shown in Fig. 6.8. This must be done as shown in Fig. 6.10: the output voltage is still measured across C_L. A good transient response, that is one showing a very small overshoot, should be obtained when R_L is about 10 Ω. The response should be the same for both positive and negative going step inputs, even when the change in the voltage across C_L is several volts. As the output transient amplitude begins to approach its maximum value, which is close to $V_+/2$ in the negative going direction but less for positive going outputs because of the voltage drop across R_9, non-linearities will be observed and these should be looked at carefully and explained.

The rise and fall times across the 1000 pF load should be in the region of 50 ns for this experimental circuit. For an output voltage of 5 V, this means that transient currents close to 100 mA are flowing in Q_3 and Q_4. The paths taken by these large transient currents, in the experimental circuit shown in Fig. 6.8, should be considered and the voltage at the emitter of Q_3, and on the positive power supply line, should be looked at with the oscilloscope on a sensitive range, using the a.c. input option. It is also interesting to look at the change in voltage which occurs at the common bases of Q_1 and Q_2. The approximate transient analysis of the previous section assumed that this was negligible.

In conclusion, the problem of designing a wide-band d-c deflection amplifier for a conventional analog oscilloscope has brought out the importance of feedback. The way feedback is supplied to the various stages of amplification is central to this design problem. In the input stages the adoption of a shunt feedback stage driving a series feedback stage, which, in turn, drives the next shunt feedback stage, means that the gain of each stage is well defined, the full bandwidth is available, and that the stages couple together with no problems because the output impedance of one is just what is needed to drive the input impedance of the next.

In the output stage, considered in the final section of this chapter, feedback is again local, applied across the output stage alone, but the circuit is now one which works at high power level because of the transient current demanded by the capacitive load. This is the reason for the interesting internal structure of the circuit, seen first in Fig. 6.7 with capacitors C_1, C_2, C_3 and C_4, and then in the experimental circuit, Fig. 6.8, with C_1, C_2 and C_3.

Notes

1 Maclean, D. J. H., *Broadband Feedback Amplifiers*, John Wiley, New York, 1982.
2 Kovács, F., *High-frequency Applications of Semi-Conductor Devices*, Elsevier, Amsterdam, 1981.
3 Carson, R. S., *High-frequency Amplifiers*, John Wiley, New York, second edition, 1982.

4 In Fig. 6.1 the curve shown for the AD3554 has been copied from Fig. 1 of Analog Devices data sheet C700-9-4/82. The curve for the CA3100 has been copied from the RCA Solid State Data Book, SSD-240B, 1982, p. 137, Fig. 3. The curve for the AD5539 has been copied from Fig. 13 of Analog Devices data sheet C1044-2/87.

5 Von Ardenne, M., *Wireless Engineer*, **13**, 59–64, 1936. This paper describes a 0.2 Hz–3 MHz amplifier with a gain of 66 db. This was an outstanding achievement for the time. A more detailed picture of the state of electronics at that time may be obtained from O. S. Puckle's translation of von Ardenne's *Fernsehenempfang*, published by Chapman and Hall, London, 1936, under the title *Television Reception*.

6 Gray, P. R., and Meyer, R. G., *Analysis and Design of Analog Integrated Circuits*, John Wiley, New York, second edition, 1984, pp. 510–15.

7 Cherry, E. M., and Hooper, D. E., *Proc. IEE*, **110**, 375–89, 1963. This paper, entitled 'The design of wide-band transistor feedback amplifiers', marked a major step forward in technique. The rather sharp exchange of correspondence between P. J. Beneteau and the authors, on p. 1617 of the same volume, should be noted, however.

8 Slaughter, D. W., *Electronics* **28**, No. 5, 174–5, May, 1955.

9 The Hitachi V-1065/V, 100 MHz oscilloscope is an interesting example. This instrument was introduced around 1984.

10 Faulkner, D. W., *IEEE J. Sol. St. Circ.*, **SC-18**, 333–40, 1983. This paper, 'A wide-band limiting amplifier for optical fibre repeaters', describes a 470 MHz bandwidth, 60 db gain, amplifier which is made up of three shunt/series feedback pairs in cascade. The high frequency analysis and circuit simulation are considered in depth.

11 For data on conventional cathode ray tubes, the book by J. Czech, *Oscilloscope Measuring Techniques*, Philips Technical Library, Eindhoven, revised English edition, 1965 is still valuable. The input capacitance to the Y deflection plates of a modern CRT, specially designed for a wide-band oscilloscope with the connections to the plates available at the side of the tube, can be as small as 3 pF. The current involved would still be about 100 mA, however, because the rise time would be well below 10 ns. For today's CRT technology there is an excellent review, in English, by W. M. P. Zeppenfeld, 'Oscilloscope Tubes: past, present and future', in *Acta Electron. (France)*, **24**, No. 4, 309–16, 1981–2, which gives many references.

12 Van Schaik, H., and Zeppenfeld, K., *Electronic Engineering*, **58**, No. 710, 47–9, February 1986.

13 These simple equations for the bipolar transistor are invaluable in circuit design because they provide a simple algebraic model for the active device. In contrast to the accurate model which must be used for numerical analysis and circuit optimisation, an algebraic model is needed when the designer is still searching for a good circuit shape. The excellent text, *Solid State Electronic Devices*, by B. G. Streetman, Prentice Hall, Englewood Cliffs, NJ, second edition, 1980, deals with the charge control model on pages 259–60 and 276–8. A very valuable review is 'Past and present of the charge control concept in the characterisation of the bipolar transistor', by J. te Winkel, *Adv. Electr. Electr. Phys.*, **39**, 253–89, 1975.

7
Waveform generator circuits

7.1 Introduction

Waveform generators make up a group of instruments which are essential to the electronic circuit designer. At the simplest level, the sine wave, square wave and triangle waveform generator, covering the frequency range from a few hertz to several megahertz, is used to measure the gain and frequency response of amplifier circuits, and as a basic timing or input signal to the kind of experimental circuits that have been discussed so far in this book. Pulse generators are also of great value for circuit testing, providing both positive and negative going pulses with very fast rise and fall times, together with the facility of a separate trigger pulse from the instrument which precedes the main output pulse by some time which may be varied. A recent book by Chiang [1] covers many of these classical topics in detail and also deals with quite advanced instrumentation and signal processing techniques that rely heavily upon waveform generation.

In recent years, laboratory instruments using digital techniques for very complex waveform generation have been introduced. In these instruments, a microprocessor is used to generate the required waveform as a continuously changing digital output of eight, or more, bits, and this digital output is then converted into the required analog output by a digital to analog converter (DAC). A striking example of such an instrument is the Hewlett–Packard 8175A Dual Arbitrary Waveform Generator [2] which can generate two quite separate outputs, or, of course, related outputs, with any waveform the user needs, subject only to the limitation that these are, in fact, made up of discrete points, of ten bit accuracy, in the voltage–time plane, and that only 50 million points per second can be generated.

In connection with such digital instruments it must always be remembered that the process of digital to analog conversion may, in itself, call for a conventional waveform generator circuit. Precision ramp generators are found in the pick-off type of DAC, and are central to ADCs as well [3]. All digital instruments require accurate clock waveform generators, which call for the generation of an accurate square waveform from a crystal oscillator, and this is by no means a trivial problem. Many digital instruments may call for highly sophisticated interpolation techniques which are realised in hardware and involve precision ramp waveforms [4].

Coming back to the more conventional kind of signal generator that is needed in the electronic circuits laboratory, another strong influence in recent times has been the development of frequency synthesis [5]. This is essentially a digital technique which makes it possible to generate waveforms of very accurate frequency and then vary this frequency rapidly from one value to another. Neighbouring values of frequency may be very close indeed, however. When these techniques of frequency synthesis are combined with waveform generator circuit techniques, very remarkable instrument performance may be achieved. For example, Danielson and Froseth [6] have described an instrument that covers the frequency range 1 μHz–21 MHz, and gives sine, square, triangle and ramp waveform outputs, all of which may be modulated, or swept in frequency.

7.2 The choice of a fundamental waveform

The simplest signal source that may be found in the electronic circuits laboratory is the function generator. This can give a sine, square, or triangle waveform output. To build such an instrument a decision must be taken on which of these three waveforms should be the fundamental one: the waveform from which all the others will be derived.

A sine wave might be chosen as the fundamental waveform if *frequency* were the output variable of prime concern. This would be because one or more crystal oscillators would be used, and these naturally produce a sine wave output when the oscillator circuit is designed for frequency precision [7].

A square wave might be chosen as the fundamental waveform when *time* was the main interest for measurement. This would follow because very fast rise and fall times would be called for from the square wave, or pulse, output from the instrument so that time measurements could be made accurately between one fast transition and another. Similarly, rise and fall times of the response of some circuit under test can only be made if the input waveform has a much faster rise or fall time.

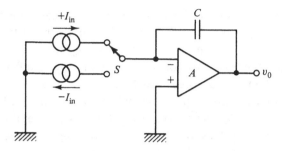

Fig. 7.1. *The basic triangle waveform generator circuit.*

A triangle wave might be chosen as the fundamental waveform when *rate of change* was the variable of prime concern. This would be the case when other waveforms, such as sine waves and other trigonometric waveforms, need to be generated with precision over a wide frequency range. The triangle waveform is thus the waveform usually found to be the fundamental one in a function generator. Very similar circuit techniques to triangle waveform generation are also found in ramp waveform generators and in the dual slope kind of triangle waveform generators which are so important in analog to digital conversion and in some interpolation techniques.

7.3 Triangle waveform generation

The generation of a triangle waveform usually depends upon one of the most elementary relationships of electronics:

$$i = C\mathrm{d}v/\mathrm{d}t. \tag{7.1}$$

Equation (7.1) describes the increase in the voltage, v, across a capacitor, C, when a current, i, is flowing into this capacitor. It follows that a really linear voltage ramp, with a constant $\mathrm{d}v/\mathrm{d}t$, calls for a really constant current source.

It is this provision of a really constant current source which is central to the triangle waveform generation problem. In order to produce a linear increase in v, followed by a linear decrease, it is necessary to have constant current sources of both signs and to be able to switch rapidly from one to the other.

Fig. 7.1 shows a first step towards a circuit shape for a triangle waveform generator. Two constant current generators are shown, $+I_{\mathrm{in}}$ and $-I_{\mathrm{in}}$, which may be switched alternately to feed the virtual earth of an operational amplifier which is connected as an integrator by means of the feedback capacitor, C. When $+I_{in}$ flows into the virtual earth, the output voltage, v_o, will fall linearly at a rate $dv_o/dt = -I_{in}/C$. When the switch, S, is in the other position, and current I_{in} is pulled from the virtual earth, the output voltage rises linearly at a rate $dv_o/dt = +I_{in}/C$.

Fig. 7.1 does not show how the switch, S, is controlled. This must obviously be done by monitoring the output level, v_o, and changing the position of S when v_o falls to some defined level. Similarly, the switch must be returned to the position shown in Fig. 7.1 when the output subsequently rises to some level. A circuit which will provide precisely such a level monitoring and switching function has already been considered in this book: Fig. 4.6. This showed an operational amplifier with positive feedback acting as a comparator circuit and having the double valued output/input characteristic shown in Fig. 4.8. The easiest way of seeing how this circuit can be added to the basic circuit of Fig. 7.1, and so form a working triangle waveform generator, is to go at once to the first experimental circuit for this chapter.

7.4 An experimental triangle waveform generator

Fig. 7.2 shows the experimental triangle waveform generator. The constant current generators and switch, shown first in Fig. 7.1, are realised by the OP07 and the CA3019 diode array, together with their associated components. The integrator of Fig. 7.1 turns up in Fig. 7.2 as the OP27 with feedback capacitor C_1. The output monitoring and switch control part of the circuit is shown at the bottom of Fig. 7.2, and involves the wide-band operational amplifier, CA3100.

The action of the CA3100 part of the circuit shown in Fig. 7.2 is as follows. Positive feedback is applied across the CA3100 via R_7 and R_8, so that, when no current flows in R_9, the output of the CA3100 must be saturated at its maximum possible level, either positive or negative. The voltage at the junction of R_7 and R_8 is limited by two 4.7 V Zener diodes which are connected back to back. This connection gives a defined voltage level of ± 5.3 V: 4.7 V plus the 0.6 V of a forward biassed silicon diode. This means that when a voltage is applied to this part of the circuit via R_9, and R_9 is made identical to R_8, the output of the CA3100 will change sign only when the output voltage of the whole circuit, that is the output from the OP27, reaches ± 5.3 V. When v_o reaches $+5.3$ V, the voltage at the junction of R_6, R_7 and R_8 will flip from -5.3 V to $+5.3$ V. When v_o falls to -5.3 V, the voltage at this junction will flip back to -5.3 V. Naturally, other critical levels may be set for v_o by changing the relative values of R_8

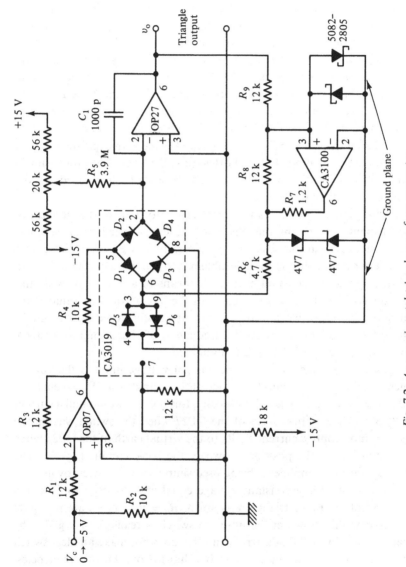

Fig. 7.2. An experimental triangle waveform generator.

and R_9. The Zener diode voltage only needs to be well below the maximum output level obtainable from a CA3100.

The CA3100 operational amplifier is thus working as a fast comparator in this circuit, a function which it performs very well indeed. It has the advantage over integrated circuits which are specially designed as fast comparators in that it naturally provides an output voltage well above or well below ground potential. It is, however, a very wide-bandwidth operational amplifier, discussed briefly in chapter 6 in connection with Fig. 6.1, and this means that this part of the circuit should be laid out very compactly with the CA3100 mounted, preferably without a holder, over a ground plane [8]. The CA3100's ± 15 V power supply pins, not shown in Fig. 7.2, must be decoupled to the ground plane with capacitors of very low inductance [9] and with the shortest possible leads. The power supply to the OP07 and OP27, also ± 15 V, must be similarly decoupled as close to the devices as possible.

Turning now to the constant current generator part of the circuit, this is implemented by noting that the constant current needs to be supplied to a virtual earth in the basic circuit first shown in Fig. 7.1. This means that a very simple solution to the problem is to provide a constant voltage, and then define the constant current by means of a resistor between this constant voltage and the virtual earth of the integrator. The value of the constant current produced in this way is then easily varied by varying the constant voltage. In this way the frequency of the triangle waveform generator may be varied or even modulated.

Fig. 7.2 shows this solution. The control voltage input is the voltage which sets the instantaneous frequency of the triangle waveform generator and is a *negative* voltage level, V_c. Ideally, this voltage would pull a current V_c/R_2 out of the virtual earth of the OP27 when the switch was in one position and supply a current V_c/R_4 to the virtual earth of the OP27 when the switch was in the position shown in the basic circuit, Fig. 7.1. The OP07 operational amplifier is being used simply as an inverter, by making $R_1 = R_3$, to provide a constant voltage equal and opposite to V_c at the input end of R_4. For a triangle waveform, $R_2 = R_4$. Things are not quite so simple as this, however, because the switch is realised, in Fig. 7.2, by means of the CA3019 diode array, and this does not make an ideal switch with negligible on-resistance and zero voltage drop. At low frequencies, when the current in R_2 and R_4 is small, the diode switch presents an on-resistance that begins to compare with R_2 and R_4, and this switch resistance is not very well defined. This is why R_5 has been connected to pin 2 of the OP27 in Fig. 7.2. The voltage at the top of R_5 is adjusted to compensate for any asymmetry in the triangle wave, which may be

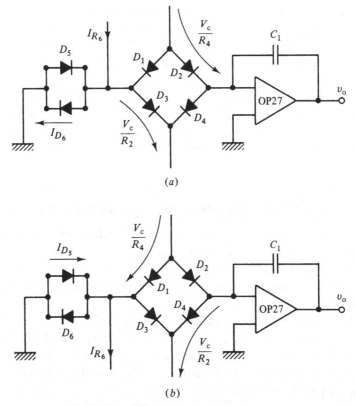

(a)

(b)

Fig. 7.3. *The two positions of the diode switch which is used in the experimental triangle waveform generator.*

observed at the lowest frequencies of operation, due to any difference in the on-resistances of D_2 and D_4 at very low currents. R_5 also supplies the small input bias current needed by the OP27. This may be of either sign and is only a few nanoamps.

The detailed functioning of the diode array as a switch is shown in Fig. 7.3. Fig. 7.3(a) shows the switch in the position first shown in Fig. 7.1, where current is supplied to the virtual earth of the integrator. In Fig. 7.2, this is the state when v_o is falling towards its lower limit of -5.3 V and the output of the CA3100 is saturated at its maximum positive level. The junction of R_6, R_7 and R_8 is at $+5.3$ V and a current, I_{R_6} will be supplied to the junction of D_1, D_3, D_5 and D_6. As shown in Fig. 7.3(a), I_{R_6} can only flow on, part through D_6, to ground, and part through D_3, where this part must equal V_c/R_2. For this to happen, R_6 must be chosen so that I_{R_6} exceeds V_c/R_2. The result of this, as may be seen by combining Fig. 7.3(a) and Fig. 7.2, is that both D_1 and D_4 have close to zero volts across them, while D_2 is forward biassed and supplying the current V_c/R_4 to the virtual earth of the OP27. This state of affairs persists until v_o falls to -5.3 V, whereupon the output from the CA3100 changes from positive to negative, I_{R_8} reverses, and the state of the diode array goes from that shown in Fig. 7.3(a) to that shown in Fig. 7.3(b).

In Fig. 7.3(b) the current I_{R_8} can only flow through D_5 and D_1. In this state, it is D_3 and D_2, in Fig. 7.2, that have close to zero volts across them and D_4 now allows the current V_c/R_2 to be drawn from the virtual earth of the OP27. The output, v_o, of the OP27 thus rises towards $+5.3$ V, and, upon reaching this, the cycle repeats.

7.5 Measurements on the experimental circuit

Fig. 7.2 shows a choice of 1000 pF for C_1, which gives a maximum frequency of operation of 20 kHz for the triangle waveform generator when the input control voltage, V_c, is at about -5 V. This integrating capacitor, C_1, must, of course, be a plastic film capacitor with very low leakage.

The most revealing experiment to be made on the circuit shown in Fig. 7.2 is a measurement of its performance as a voltage controlled oscillator: the relationship between output frequency and control voltage input, V_c. This reveals the limit in the magnitude of V_c, where the current in R_8 is no longer big enough to keep the diode array switch on, and also reveals the non-linearity in the frequency–voltage characteristic due to the imperfections of the diode array as a switch.

The limit in the magnitude of V_c is best observed by looking at the voltage on pin 5, or pin 8, of the CA3019. This should be a rectangular waveform which, in the case of pin 5, is at zero during the positive slope of the triangle wave output, and at about $+0.6$ V during the negative slope of the output. As V_c is increased in magnitude, a point will be reached where V_c/R_2, referring now to Fig. 7.3(a), equals I_{R_8} and there is not enough current left to keep D_6 conducting. The rectangular waveform at pin 8 will then be seen to leave zero level during the negative slope of the output waveform: D_4 is no longer being turned off when it should be. The same kind of thing may be seen at pin 5 during the positive slope of the output waveform.

The non-linearity in the frequency–voltage characteristic is shown in Fig. 7.4, where the experimentally determined variation in f against the magnitude of V_c is shown, for the component values shown in Fig. 7.2. A large non-linearity can be seen for values of $|V_c|$ approaching the forward drop across the diodes, D_2 and D_4. Nevertheless, a whole decade of

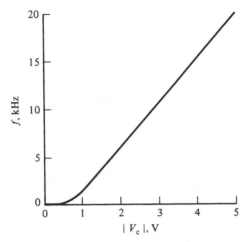

Fig. 7.4. *The voltage–frequency characteristic of the experimental triangle waveform generator.*

frequency, 2–20 kHz, may be covered with this circuit, and the frequency–voltage characteristic is very linear over this decade. By increasing the value of C_1 in decades, a wide range of frequency may be covered. If C_1 is made less than 1000 pF, on the other hand, slew rate limiting will eventually be observed even though the OP27 is quite a fast precision amplifier. The generation of precision triangle waveforms at megahertz frequencies is not possible using simple integrated circuits: special techniques will be called for, and these will be considered towards the end of this chapter.

7.6 A better method of output level control

A rather crude feature of the experimental circuit shown in Fig. 7.2 is the very simple way in which the amplitude of the output is set by means of the two back to back Zener diodes. This must introduce a slight temperature dependence of the amplitude of the output, which may be minimised by choosing a Zener diode that has a Zener voltage temperature dependence equal and opposite to the temperature dependence of the voltage drop across the same type of diode when in forward conduction. This choice may not, of course, be the one which would also give the required output level and, in any case, this does not help with another problem which is the fact that the current through the two Zener diodes does vary slightly as V_c is varied.

The circuit may be improved if the two Zener diodes, in Fig. 7.2, are

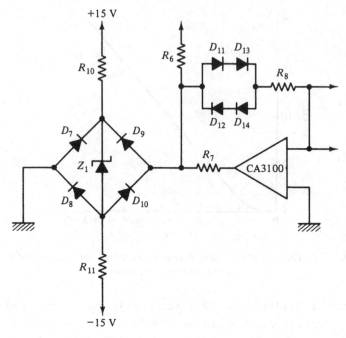

Fig. 7.5. An improvement to the experimental circuit shown in Fig. 7.2,
which replaces the two Zener diodes shown there.

replaced with the rather complicated, and more expensive, connection of a precision voltage reference [10], eight diodes, and two additional resistors, R_{10} and R_{11}, as shown in Fig. 7.5. This voltage level control is far better at high frequency too, because the voltage across the voltage reference, Z_1, is always in the same direction, whereas the two Zener diodes, in Fig. 7.2, must continuously cycle between forward conduction and reverse breakdown.

Note that the forward drop across D_8 and D_9, which is, of course, temperature dependent, should be compensated by the forward drop across D_{11} and D_{13}, while D_{12} and D_{14} should compensate for the forward drop across D_7 and D_{10}. For this to happen, the forward currents of all these diodes need to be about the same, regardless of the current in R_6. This is not easy to arrange when a diode array switch is being used, because the current in R_6 will be quite large. A combination of the level control shown in Fig. 7.5 with an FET switch gets around this problem, however, and the diode currents may be made about equal by connecting a resistor from the junction of D_{13}, D_{14} and R_8 to ground [11].

7.7 Triangle to sine wave conversion

The important feature of the triangle waveform is its precisely linear rate of rise and fall. This may be exploited when some other waveform is needed as an output from an instrument which is, fundamentally, a triangle waveform generator. It is only necessary to design a non-linear circuit which has an input–output characteristic of the same shape as the required waveform, and ensure that this non-linear characteristic is independent of frequency over the required range.

A classical solution to this problem [12] is to use diodes to switch the feedback resistors across an operational amplifier, and thus change the gain as the output, or input, voltage changes. This solution is still found in quite recent instruments to solve the triangle to sine wave conversion problem [13], but it is essentially a piece-wise linear approximation and this means unavoidable discontinuities in the slope of the output waveform.

A far better solution is to use a smooth non-linearity, like the output–input characteristic of the long tailed pair circuit: the one shown in Fig. 4.5. This approach has been reviewed in an important paper by Gilbert [14], who gives a number of references to earlier work on the problem of obtaining an output–input characteristic that is an accurate match to the sine function over the range $-\pi/2$ to $+\pi/s$ in its argument.

The main problem is reproducing the zero slope that the sine function has at its maxima and minima. As Fig. 4.5 shows, the simple long tailed pair output–input characteristic approaches zero slope only asymptotically. It was, perhaps, Gilbert who first proposed the solution to this problem with a circuit of the interesting shape shown in Fig. 7.6 [15]. This effectively superimposes a number of characteristics of the kind shown in Fig. 4.5, which, alternately, have opposite sign.

As shown in Fig. 7.6, the input voltage, v_{in}, is applied directly to the long tailed pair at the centre of the circuit, Q_3 and Q_4. Referring to equations (4.1)–(4.7), it follows that Q_3 and Q_4, taken on their own, will produce a difference in output currents,

$$(i_1 - i_2)/I_L = \tanh(v_{in}/2V_T) \tag{7.2}$$

where $V_T = kT/e$, and is close to 25 mV at room temperature.

The bias current, I_B in Fig. 7.6, is arranged to produce an offset voltage,

$$V_B = I_B R \tag{7.3}$$

where R is the value of the resistors shown in Fig. 7.6. The currents in Q_1 and Q_2 are then equal when $v_{in} = +V_B$, and the currents in Q_5 and Q_6 are

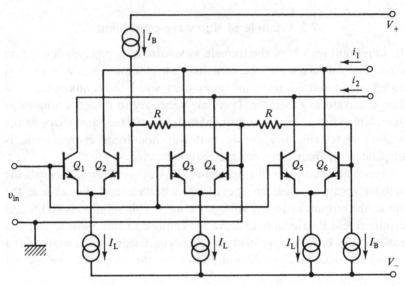

Fig. 7.6. *A circuit shape which may be used to convert a triangle waveform into a sine wave.*

equal when $v_{in} = -V_B$. It follows that the complete circuit shown in Fig. 7.6 has the transfer function

$$(i_1 - i_2)/I_L = \tanh(v_{in}/2V_T) - \tanh[(v_{in} - V_B)/2V_T] - \tanh[(v_{in} + V_B)/2V_T].$$
$$(7.4)$$

A judicious choice of V_B can make the function given by equation (7.4) agree with the sine function to within a few per cent over the range of $v_{in}/2V_T$ which would then be used to correspond to $\pm\pi/2$. This is shown in Fig. 7.7, where equation (7.4) has been plotted for the particular case of $V_B = 75\,\text{mV}$, assuming that $V_T = 25\,\text{mV}$. The function shows a maximum, $+0.2923$, occurring at $v_{in}/2V_T = 0.75$, and a minimum, -0.2923, occurring at $v_{in}/2V_T = -0.75$. The function also shows an asymptotic approach to ± 1 as $v_{in}/2V_T$ tends to $\mp\infty$.

By driving the circuit shown in Fig. 7.6 with a triangle waveform of amplitude $1.5\,V_T$, this amplitude being kept strictly constant as the frequency is varied, an output $(i_1 - i_2)$ will be obtained that is a very close match to a sine wave with an amplitude of $0.2923I_L$.

Gilbert's later developments of circuits along these lines [14], which he had termed 'translinear circuits' in an earlier work [16], led to remarkably powerful monolithic circuits which could synthesise all the standard trigonometric functions, and their inverse functions, over a much wider range of argument than the limit of $\pm\pi/2$ which has been put forward

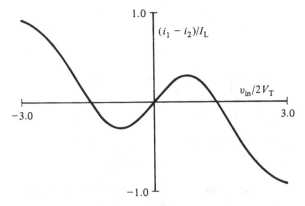

Fig. 7.7. *The output–input characteristic of the circuit shown in Fig. 7.6 when $I_B R$ is made equal to 75 mV.*

here [17]. These provide a solution to the triangle to sine wave conversion problem and are also the solution to many other complex waveform generation problems, even up to frequencies around 10 MHz.

7.8 Problems in fast waveform generation

So far in this chapter, only fairly low frequency waveform generators have been discussed. Even so, the experimental triangle waveform generator of Fig. 7.2 was limited in its performance by the slew rate of the operational amplifier which was used as an integrator. It is clear that the generation of triangle and ramp waveforms with frequencies well up in the 10 MHz region, or above, will call for special techniques.

Fast ramp waveforms are needed for the timebase in simple analog oscilloscopes. A 100 MHz analog oscilloscope, for example, will certainly need a timebase waveform fast enough to sweep one screen diameter in 100 ns. The actual waveform generating circuit would operate at quite low level, perhaps giving a 5 V sweep in 100 ns, and then the problem of generating the waveform which is needed to drive the CRT is shifted onto the design of the deflection amplifier, a topic which was considered in section 6.11. Nevertheless, the 5 V in 100 ns, mentioned above, is a rate of change of 50 V/μs, and this calls for special techniques.

Fast pulse waveform generation has already been considered in this book in chapter 2. There, the techniques used to generate the very short pulses needed for high speed sampling gates were discussed, and, as with most high speed switching circuits, attention was focused on the device being used: the step recovery diode. Some fast pulse and square wave generation problems are more circuit conscious, however, and an example

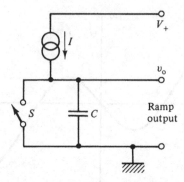

Fig. 7.8. A beginning to the solution of the fast ramp generation problem.

is the classical sine wave to fast square wave conversion problem. This was mentioned above in section 7.1, and will be the final topic of this chapter. Before that, the fast ramp or timebase waveform problem will be considered.

7.9 Fast ramp generation

The simple circuit shape shown at the beginning of this chapter, Fig. 7.1, fails when applied for very fast ramp waveform generation for two reasons.

The first reason is the limitation on the slew rate at the output of the operational amplifier. This must always be much greater than the maximum rate of change of voltage which is called for at the output, otherwise the amplifier is no longer working as a linear device.

The second limitation on the simple circuit shape shown in Fig. 7.1 is the finite propagation delay across the operational amplifier. If the switch, S, is to change position very rapidly, to produce the transition in the waveform from the positive going ramp to the negative going ramp by reversing the current flow into the virtual earth, the same rapid reversal should take place in the current flowing in the feedback capacitor, C. However, this reversal will not take place instantaneously because of the delay through the amplifier, A. This will mean that the input circuit of the operational amplifier will become saturated and a finite time will be needed for its recovery.

It follows that fast ramp waveform generation is usually tackled with a quite different circuit shape to the one shown in Fig. 7.1.

A first step towards a new circuit shape is shown in Fig. 7.8. A capacitor, C, is shown, charging from a constant current source, I, and thus producing a linear sweep, $dv_o/dt = I/C$. The sweep is initiated by

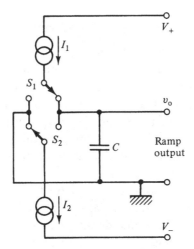

Fig. 7.9. Fast dual-slope ramp generation.

opening the switch, S, very rapidly, so that the current, I, is simply diverted, from flowing through S to ground, into the capacitor. The output voltage, v_o, rises linearly to some pre-determined level, where a level detecting circuit would send a control signal back to the switch, S, and make it close again. This would generate the fast resetting of the ramp output voltage: the flyback of a timebase waveform, for example. If an accurate dual-slope ramp waveform were needed, as it is in many ADC applications [3] and in the digitising oscilloscope fine interpolator, mentioned at the beginning of this chapter [4], two change-over switches would have to be used, as shown in Fig. 7.9, with two separate constant current sources, I_1 and I_2.

An important feature of both circuits, shown in Figs. 7.8 and 7.9, is that the constant current generators will always be working into a well-defined load, provided the switches are make before break switches. This is in contrast to the simple integrator idea of Fig. 7.1, where the finite propagation delay in the operational amplifier meant that the constant current generators would find themselves loaded by a saturated input circuit, instead of a true virtual earth, every time the switch, S, changed position. In fact, this problem may be seen experimentally with the circuit, Fig. 7.2, which was used earlier in this chapter. A sufficiently sensitive wide-band oscilloscope will show that the virtual earth of the OP27, used there as an integrator, is no true virtual earth at all for the fraction of a microsecond during which the triangle output waveform is changing the sign of its slope. On the rather slow time scale of this experimental circuit, however, this is no problem.

7.10 A practical circuit

Fig. 7.10 shows the practical realisation of the kind of circuit shown here first as Fig. 7.8. The constant current generator, I in Fig. 7.8, is realised in Fig. 7.10 by means of Q_1, which is in grounded base configuration with its base returned to a well-defined voltage level a few volts below V_+. This voltage level is defined by the Zener diode, D_1, while the normal silicon diode, D_2, is included to make some compensation for the temperature dependent V_{BE} of Q_1. In this way the current in Q_1 is defined at V_{D_1}/R_1, where V_{D_1} is the Zener voltage of D_1.

Q_2, in Fig. 7.10, plays the part of the make before break switch, S, in Fig. 7.8. The trigger input simply turns Q_2 either off, to initiate the ramp, or on, to terminate the ramp and discharge C_1 ready for the next sweep of the timebase. This trigger input, in a practical situation, would come from the control circuits of the oscilloscope. The negative going transition of the trigger input, which turns Q_2 off, would come from the timebase trigger circuit, while the positive going edge, which turns Q_2 on again, would be generated by some level comparator circuit in the timebase output amplifier, which would detect the fact that the CRT display had swept out a full screen diameter.

For an analog oscilloscope, the fairly simple kind of constant current source shown in Fig. 7.10 will be good enough. The timebase range will be set by switching values of R_1 and C_1; Q_2 needs to be a transistor that will switch rapidly, to minimise the delay in starting the sweep once the trigger edge has been generated. Q_2 must also have a very low leakage current when it is turned off. Similarly, the wide-band amplifier which follows this circuit should have a very high input impedance, so that negligible current is drawn from the capacitor, C_1, as the voltage across it builds up. If this were to happen the linearity of the sweep would be degraded.

The temperature dependence of the constant current source shown in Fig. 7.10 is acceptable for simple analog oscilloscope applications because a fine adjustment is always provided so that the timebase speed can be set, against the oscilloscope graticule, using an accurate internal clock waveform. The sources of temperature dependence, in the circuit shown in Fig. 7.10, are the variation of V_{BE} for Q_1, which can never be perfectly matched by the simple diode, D_2, and the variation of h_{FE} in Q_1, which will change the base current flowing from Q_1, this being subtracted from the defined current, V_{D_1}/R_1, to leave the current which actually defines the sweep speed. There is also the temperature dependence of V_{D_1} itself.

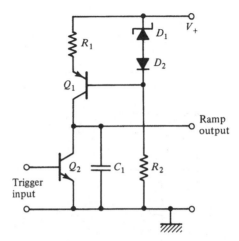

Fig. 7.10. *The kind of circuit shape found in an analog oscilloscope for the timebase generator.*

7.11 Precision ramp generation

When a very precise timebase waveform is called for, a much better kind of current source circuit must be developed, compared to the simple single transistor circuit shown in Fig. 7.10.

Fig. 7.11 shows the kind of circuit shape that is needed. The current source is now provided by means of a precision voltage reference, Z_1 (the same device dealt with in section 7.6 [10]), and an instrumentation operational amplifier, A_1, with a very low input offset voltage and input bias current. In this new circuit the reference voltage, V_{z_1}, is accurately reproduced across the resistor R_1. Furthermore, the use of an MOST for Q_1, in place of the bipolar device previously used in Fig. 7.10, means that there is no temperature dependent current subtracted from the defined current, V_{z_1}/R_1.

The use of a p-channel enhancement MOST for Q_1, in the circuit shown in Fig. 7.11, has the great advantage of putting the output voltage level of A_1 several volts below the positive rail, V_+, so that A_1 can be powered from this same positive supply.

The MOST should also have a more well-defined output capacitance, compared to the bipolar transistor of the previous circuit, and this brings up two more points which should be considered when precision ramp generators are being designed.

The first point concerns factors which may make the value of C_1 change as the voltage across it builds up. The dielectric used in C_1 must be free from any non-linearities of this kind. Ceramic dielectric capacitors are, of

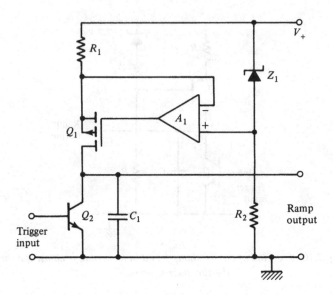

Fig. 7.11. Schematic of a timebase waveform generator circuit using a precision current source.

course, quite unusable in this respect, but it is interesting to note that many very complex non-linear and time dependent effects remain to complicate the designer's work, even when the highest quality capacitors are used [18].

The second point concerns the output capacitances of Q_1 and Q_2. These will both be in parallel with the capacitor C_1 but, because Q_1 and Q_2 will both be high frequency devices, the additional capacitance should be only a few picofarads. Nevertheless, this will impose a lower limit on C_1 because the small voltage dependent part of the output capacitances of Q_1 and Q_2 must be kept negligible in comparison to C_1. This means a lower limit of about 50 pF for C_1, and that really high speed ramp generation must call for quite high currents.

Finally, the output capacitance of Q_1 may introduce problems at very high speed because a small high frequency current can be supplied to the output of A_1 through this capacitance, and the output impedance of A_1 may well be quite high at high frequency. This problem may be overcome by using a second p-channel MOST in cascode connection with Q_1. The gate of this second device will be taken to a well-decoupled bias point, and thus will isolate A_1 completely from the fast waveform across C_1.

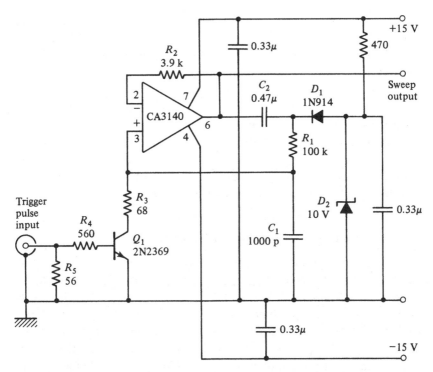

Fig. 7.12. *An experimental ramp generator circuit which illustrates bootstrapping.*

7.12 The bootstrap technique

The problem of providing a constant current source for precision ramp waveform generation can be solved by using a quite different technique to those described in the previous sections. This is the bootstrap technique, and is, in fact, a very old idea in electronic waveform generation [19].

As with the previous circuits, Figs. 7.10 and 7.11, constant current in a bootstrap circuit is produced by keeping the voltage across a resistor constant. The bootstrap circuit, however, exploits the fact that the ramp output waveform must be taken to the outside world through an amplifier that presents a very high input impedance to the timing capacitor, C_1 in the previous circuits, and also has a very low output impedance. This same low output impedance can then be used, after a level shift, to drive the current into the timing capacitor via the current defining resistor. The bootstrap technique is thus an example of positive feedback.

The technique should be made clear from Fig. 7.12, which shows the second experimental circuit for this chapter. In Fig. 7.12, the ramp waveform is being generated across the timing capacitor, C_1, which, as previously, has a fast switching transistor, Q_1, connected across it. C_1 is charged by the current in R_1, and, initially, this current is defined by the Zener diode, D_2, which sets the voltage across R_1 to be 10 V, less the forward drop across D_1 and the $V_{CE(SAT)}$ of Q_1. The resistor R_3 may be ignored at this stage.

When Q_1 turns off, C_1 begins to charge. Through C_2, which is made much larger than C_1, the voltage follower (the CA3140) pulls up the top of R_1 (by its bootstraps!), turning off the diode D_1. The voltages at the top and the bottom of R_1 both rise, both at exactly the same rate if the voltage follower gain is unity, and the ramp will be linear because the current in R_1 is being kept constant by means of this bootstrap action.

7.13 Measurements on the bootstrap circuit

The values shown in Fig. 7.12 give a ramp which rises at a rate just below 0.1 V/μs. This ramp is reset by a positive trigger pulse input which needs to be above 1 V amplitude and about 10 μs duration. The repetition rate of the trigger pulse input determines the amplitude of the ramp output, and at repetition rates below 5 kHz the ramp has time to reach the maximum possible positive level, which is a few volts below the $+15$ V supply. Note that the top of R_1 is then at a level well above the positive supply voltage.

Good linearity in the ramp output depends upon several factors in this circuit. The most important factor is that C_2 must be very large compared to C_1, because a change in voltage across C_2, during the ramp build up, will mean a departure from constant current in R_1. Other factors are leakage in Q_1 and leakage in C_1, which must be a high quality plastic film capacitor. All these factors tend to *reduce* the rate of rise of the ramp as its amplitude builds.

An interesting development of the bootstrap circuit can be tried out on the circuit shown in Fig. 7.12, which corrects for the factors listed above, and can, in fact, overcorrect for them and make the ramp *increase* its slope as its amplitude builds up. This development is to make the voltage follower, the CA3140, have a gain just above unity. This is easily arranged in Fig. 7.12 by adding another resistor, R_6, from pin 2 of the CA3140 down to ground. The resistor R_2 is only in Fig. 7.12, at the moment, to limit the current fed back to the inverting input: this is a recommendation on the CA3140 data sheet. When R_6 is added, R_2 plays a second role, and the voltage follower gain is increased from unity to $(1 + R_2/R_6)$. Making $R_6 = 39$ kΩ, for example, will give a gain of 1.1, and a slight curving *upward* of the ramp output should be observed.

The trigger pulse input part of the experimental circuit, R_4 and R_5 in Fig. 7.12, has been made particularly simple because there are some very interesting observations to be made concerning the way in which the ramp actually begins.

When the positive trigger pulse arrives, Q_1 turns on and discharges C_1. The current is limited by R_3, assuming that Q_1 saturates rapidly, and the 'flyback' of the ramp should take place in well under 1 μs when the trigger pulse amplitude is above 1 V.

It is the duration of the positive trigger pulse which is the most interesting experimental variable. When the trigger pulse ends, Q_1 should turn off rapidly. The ramp should then begin, from a small positive level, the $V_{CE(SAT)}$ of Q_1, without any discontinuity in level, only a discontinuity in slope. A discontinuity in level will be seen, however, if the trigger pulse is not long enough to allow Q_1 to recover from the rather large current it passed during flyback and settle down to being saturated and passing only the rather small current flowing in R_1. This behaviour should be observed closely, because there is much to understand here.

7.14 Sine wave to square wave conversion

The final waveform generation problem to be considered in this chapter is a very classical one: the conversion of a sine wave into a fast square wave over a wide range of frequency, and with good correspondence between the zero crossings of the sine wave and the fast edges of the square wave.

A possible solution to this problem would be to build a wide-band amplifier and make it have a well-defined limiting action at its output stage. For example, suppose the input sine wave has an amplitude of 100 mV and the output is to be a square wave of 1 V amplitude. Further suppose that the rise and fall of the square wave must take up only 1 % of the total period. It follows that an amplifier with a gain of $1000/\pi$, just over 300, will be needed. This is a gain of 50 db and, glancing back at the data given in Fig. 6.1, it is clear that a gain of 50 db over a bandwidth of more than 10 MHz is not something which is easy to arrange. When this is coupled with the need for a well-defined limiting action in the output stage of this amplifier, which will produce negligible time delay in either going into its limit or coming out of it, the circuit design problems concerned with the sine wave to square wave conversion hardware begin to look quite formidable.

It is interesting, for the above reasons, to look at what can be achieved in this area by using one of the fast comparators which are available as

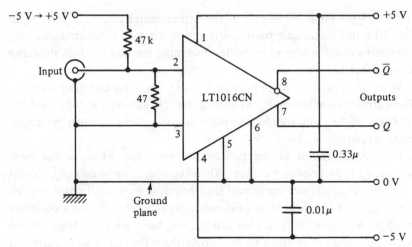

Fig. 7.13. *An experimental sine to square wave conversion circuit using a very fast comparator.*

integrated circuits. Comparator circuits were the subject of chapter 4 of this book, where the problem was one of capturing a very small input signal to the circuit at a well-defined time. Here the problem is one of amplification and limiting, but this is precisely what the simplest comparator circuits are designed to do. What is surprising about these devices is their remarkable bandwidth.

7.15 Experiments with a fast comparator

An example test circuit is shown in Fig. 7.13. This uses a fairly recent fast comparator: the LT1016 [20]. The experimental circuit shown in Fig. 7.13 is almost trivial in its simplicity, but it must be built very carefully with the LT1016 mounted over a ground plane, the two decoupling capacitors mounted with the shortest possible leads, and with the 47 Ω terminating resistor mounted as close to the device as possible. The output should be viewed with a wide bandwidth oscilloscope, using a passive 100:1 probe that loads the output with 5 kΩ.

The first test that should be made is an attempt to measure the small signal gain of the LT1016 and its small signal bandwidth. Connect an RF sinusoidal signal generator to the input and apply about 100 μV rms at 1 MHz. If the circuit has been constructed properly, it should then be possible to apply a voltage to the top of the 47 kΩ in Fig. 7.13 and bring the output level at either the Q or $\bar{Q}$ output, to a stable 1 V mean level. This is close to the output level at which the LT1016 should exhibit maximum small signal gain, and this gain should be about 70 db.

What is, perhaps, very unexpected about the LT1016, and, of course, other comparators of the same high quality, is the bandwidth over which this gain of 70 db is maintained. Increasing the input frequency will show an almost flat frequency response right out to around 20 MHz, where there may well be an increase in gain of several decibels before the gain collapses dramatically. Looking back at Fig. 6.1 shows that this performance is quite remarkable compared to the wide bandwidth operational amplifiers being discussed in chapter 6. In fact, 70 db is a gain of nearly 3500 so that its persistence up to 20 MHz could be interpreted as a gain-bandwidth product of 70 GHz. How is this possible with an integrated circuit fabricated using a bipolar process which gives transistors having cut-off frequencies of only a few gigahertz? The answer lies in the way in which the gain of the LT1016 rapidly collapses above 20 MHz. There is no gain at all at frequencies much above 20 MHz, and the LT1016 cannot be used with negative feedback to provide a well-defined wide-band gain.

The LT1016 has an output level limited between a few hundred millivolts above ground and about 4 V above ground. This makes the device compatible with standard logic and the limiting action is achieved with very little time delay. There is an excellent application note on the LT1016, which has been referred to before in this chapter [8].

The output limiting, and the function of the LT1016 as a solution to the sine wave to square wave conversion problem, may be studied by simply increasing the level of the sine wave input above the 100 μV which was used to check the small signal gain. Using an input sine wave of 100 mV peak, the LT1016 produces a good square wave at 10 MHz with the rise and fall times limited by the LT1016 to about 5 ns and the zero crossing of the square wave about 10 ns behind the zero crossing of the sine wave input. This 10 ns delay is the propagation delay of the LT1016 itself.

Reducing the input frequency, keeping the amplitude of the sine wave input constant at 100 mV, shows up what might be called the gain limited performance, as opposed to the speed limited performance which has just been seen at 10 MHz. Once the frequency is low enough to allow the LT1016 to reproduce the zero crossing of the input signal by means of its true small signal gain, the rise and fall times of the output square wave become input frequency dependent. A calculation along the lines of the one given at the beginning of section 7.14 shows that the rise and fall times will take up far less than 1 % of the total period, because of the high small signal gain of the LT1016. The thing for the experimentalist to watch out for, however, is just how the circuit behaves as the input level goes through zero, because it is here that any instability will be noticed. Instability can be due to poor layout, poor component quality, or poor constructional technique.

7.16 Conclusions

The first experimental circuit in this chapter, the triangle waveform generator, operated at only audio frequencies. Despite this, a number of speed limitations could be seen: the gain and slew rate demands upon the operational amplifier being used as an integrator, the speed required from the comparator used in level control, and the speed limitations that might arise in the switching part of the circuit which brings about a reversal in the current supplied to the virtual earth of the integrator.

Output level control was discussed briefly in the section following this first experimental circuit. This is a vital factor when a triangle waveform is to be supplied to one of the triangle to sine wave conversion circuits, discussed in section 7.7, because the triangle wave amplitude must be kept constant over a very wide frequency range.

When precision waveform generation is called for, coupled with a need for high speed as well, the circuit design problems can get really difficult. Precision current sources, of the kind shown in Fig. 7.11, will solve precision waveform problems at low frequencies. Bootstrap techniques are more likely to give a solution at high speed, and a good illustration of this is the precision dual-slope DAC described by Mack, Horowitz and Blauschild [21]. In the experimental bootstrap circuit discussed in this chapter, Fig. 7.12, fine tuning of the ramp linearity could be illustrated by reducing the feedback across the voltage follower used there.

The chapter ended with an experiment that was rather unusual in comparison to the other experimental circuits in this book: the circuit detail was glossed over by treating the LT1016 fast comparator as a 'black box'. The circuit detail of the LT1016 is given, perhaps in a somewhat idealised form, in the data sheet [20], and it is well worth while spending some time understanding it.

Notes

1 Chiang, H. H., *Waveforming and Processing Circuits*, John Wiley, New York, 1986.
2 Neumann, U., Vogt, M., Brilhaus, F., and Husfeld, F., *Hewlett–Packard J.*, **38**, No. 4, 4–12, April 1987.
3 Gordon, B. M., *IEEE Trans. Circ. Syst.*, **CAS-25**, 396 and 405, 1978.
4 Rush, K., and Oldfield, D. J., *Hewlett–Packard J.*, **37**, No. 4, 7–8, April 1986.
5 Two excellent books almost cover this field: *Frequency Synthesizers: Theory and Design*, by V. Manassewitsch, John Wiley, New York, third edition, 1987, and *Frequency Synthesis by Phase Lock*, by W. F. Egan, John Wiley, New York, 1981.

6 Danielson, D. D., and Froseth, S. E., *Hewlett–Packard J.*, **30**, No. 1, 18–26, Jan 1979.

7 O'Dell, T. H., *Electronic Circuit Design: Art and Practice*, Cambridge University Press, Cambridge, 1988, pp. 107–10.

8 Williams, J. M., High Speed Comparator Techniques, in: *Linear Applications Handbook*, Linear Technology Corp., Milpitas, California, 1986, pp. 13.1–13.32.

9 These would be multilayer ceramic capacitors, between 0.1 μF and 0.47 μF, of the lowest possible voltage rating in order to get the smallest possible size. The dielectric would be X7R or Z5U.

10 Examples of such devices are the 5 V REF50Z, and 2.5 V REF25Z, made by Plessey; the 6.9 V LM329, made by National Semiconductor; and a number of so-called band-gap references which all give a reference voltage close to 1.2 V. All these devices work over a much wider current range, and have a temperature coefficient several orders of magnitude better, than simple Zener diodes.

11 O'Dell, T. H., *Electronic Engineering*, **61**, No. 753, 28 September 1989.

12 Ritchie, C. C., and Young, R. W., *Electronic Engineering*, **31**, 347–51, June 1959.

13 For example, the Hewlett–Packard 3312A Function Generator uses a 12 diode shaper circuit, which gives less than 0.5% distortion below 50 kHz. Details of this shaper are given by R. J. Riedel and D. D. Danielson, *Hewlett–Packard J.*, **26**, No. 7, 18–24, March 1975.

14 Gilbert, B., *IEEE J. Sol. St. Circ.*, **SC-17**, 1179–91, 1982.

15 Gilbert, B., *Electronics Letters*, **13**, 506–8, 1977.

16 Gilbert, B., *Electronics Letters*, **11**, 14–16 and 136, 1975.

17 The AD639 Universal Trigonometric Function Generator from Analog Devices is an example. This can provide an output proportional to $[\sin(x_1 - x_2)]/[\sin(y_1 - y_2)]$, where x_1, x_2, y_1 and y_2 are all signal inputs. In the simple triangle to sine wave conversion application, the AD639 can work quite accurately over an input range corresponding to $\pm 5\pi/2$, and may, therefore, be used as a frequency doubler or tripler into the bargain. The device is described on pages 3.123–3.134 of the *1986 Update and Selection Guide*, published by Analog Devices, Norwood, Massachusetts.

18 Buchanan, J. E., *IEEE Trans. Instr. Meas.*, **IM-24**, 33–9, 1975.

19 Ridenour, L. N. (Ed.), *Waveforms*, McGraw Hill, New York, 1949, pp. 267–78.

20 *1986 Linear Databook*, Linear Technology Corp., Milpitas, California, p. 5.36.

21 Mack, W. D., Horowitz, M., and Blauschild, R. A., *IEEE J. Sol. St. Cir.*, **SC-17**, 1118–26, 1982.

8

Switched capacitor circuits

8.1 Introduction

Switched capacitor circuits, while making use of very well-established principles, appear to be a fairly recent development. The reason for this is the advance in very large scale CMOS integrated circuit technology. Switched capacitor circuits, as their name implies, are made up from switches, capacitors, and the operational amplifier circuits that are always needed to counteract the inevitable power losses and which also provide useful low impedance signal outputs. These switches, precision capacitors and good quality operational amplifiers can now all be made together within one large integrated circuit. The switches are CMOS transistors, which make excellent bilateral switches with very low on-resistance and very high off-impedance. The capacitors are MOS capacitors, which can be made in very accurate relative ratios of capacitance by controlling the area of metallisation. The operational amplifiers, which can now be made using only CMOS transistors [1], have excellent gain-bandwidth and power output.

Switched capacitor circuits can be conveniently divided into two classes. The first class exploits the concept of charge conservation. The second class could be called charge pumping circuits. These two classes are discussed in the next two sections.

8.2 Charge conservation

When a capacitor, C, is charged, so that it has a voltage, V, across it, the charge in the capacitor is

$$Q = CV. \tag{8.1}$$

This fundamental relationship suggests the possibility of effecting very

136

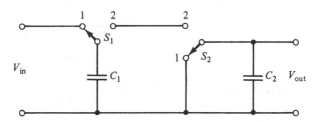

Fig. 8.1. *A switched capacitor circuit that gives voltage division.*

accurate voltage division by simply switching a charged capacitor across a second, uncharged, capacitor. The charge, Q in equation (8.1), must be conserved: it is redistributed between the two capacitors.

Fig. 8.1 illustrates this kind of circuit in its simplest form. S_1 and S_2 are break before make switches. Initially, the two switches are in position 1, so that C_1 is charged up to the voltage V_{in} and C_2 is completely discharged. The switches are then moved into position 2. Capacitors C_1 and C_2 are now in parallel and must conserve and hold the initial input charge, $C_1 V_{in}$. It follows that

$$C_1 V_{in} = (C_1 + C_2) V_{out} \qquad (8.2)$$

and that the output voltage, always assuming that this is observed with an infinite impedance voltmeter, will be

$$V_{out} = V_{in} C_1 / (C_1 + C_2). \qquad (8.3)$$

Equation (8.3) shows that a precision voltage division may be effected by ensuring that C_1 and C_2 are made in a precision *ratio* to one another: the absolute values of C_1 and C_2 are not important. Now the value of an MOS capacitor is determined by area, and it is precisely ratios in areas that can be determined with great precision with the photolithographic process which is used in the manufacture of integrated circuits. It follows that switched capacitor circuits of the charge conservation class fit naturally into the field of VLSI CMOS circuits.

The simplest voltage division operation that can be made with the circuit shown in Fig. 8.1 is division by 2: C_1 and C_2 are made identical to one another. This binary division is central to the VLSI switched capacitor ADCs and DACs, which began to be developed in the 1970s as cheap MOS devices for digital telephony [2]. The idea of making an ADC of this kind, using discrete components, was proposed as early as 1962 by Barbour [3].

There are other kinds of ADC and DAC which may be realised using switched capacitor technique. An excellent account of these may be found in chapter 7 of the text by Allen and Sánchez-Sinencio [4], who give many

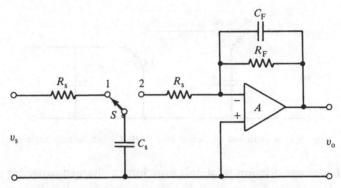

Fig. 8.2. *A charge pump circuit. Switch S is break before make and is in position 1 for time $T_s/2$, and then in position 2 for time $T_s/2$, this cycle repeating every T_s.*

references. These circuits will not be discussed further in this book, however, because it is not easy to propose interesting experimental circuits in this area which can be built easily. This does not imply that switched capacitor ADCs and DACs in VLSI CMOS are not important, they most certainly are, but circuits which use the switched capacitor technique in the charge pumping mode, the subject of the next section, allow a great variety of experimental work to be done. For this reason, this chapter concentrates on circuits of this charge pumping class.

8.3 Charge pumping

In the class of switched capacitor circuits that can be called charge pumping circuits, attention is directed on only one capacitor at a time in the circuit. This particular capacitor is being charged and then discharged, repetitively. It is the *frequency* of charging and discharging which is now one of the most important variables in the problem, whereas frequency was not mentioned at all in the previous section.

   Fig. 8.2 shows an example of a switched capacitor circuit which operates in the charge pumping mode. C_s is charged up to the signal input voltage, v_s, when the switch, S, is in position 1. The resistance which determines how fast C_s can charge up is shown as R_s in Fig. 8.2, and represents the total circuit resistance. This is made up of the source resistance, the series resistance of the capacitor, and the resistance of the switch; the latter being by far the most important when the switch is realised by means of CMOS transistors. It is assumed that $C_s R_s$ is very small compared to the time that the switch remains in position 1. The capacitor will then be charged up to the level of the input signal voltage, v_s.

The switch is, as in the previous circuit, a break before make switch. When it is moved into position 2, shown in Fig. 8.2, capacitor C_s is completely discharged into the virtual earth of the operational amplifier, A. The current pumped into the virtual earth must then be a series of exponential pulses, having the form $(v_s/R_s)\exp(-t/C_s R_s)$, occurring every T_s, where T_s is the period of the switching frequency, f_s. The mean current flowing into the virtual earth will then be

$$\bar{i} = (1/T_s)\int_0^{T_s}(v_s/R_s)\exp(-t/C_s R_s)\,\mathrm{d}t. \qquad (8.4)$$

Because the assumption $C_s R_s \ll T_s$ is essential to the operation of any charge pumping circuit, the upper limit of integration in equation (8.4) may be replaced by infinity. The mean current is then given by the very simple result,

$$\bar{i} = C_s v_s/T_s \qquad (8.5)$$

which shows up the charge pumping concept very clearly: every T_s a charge $C_s v_s$ is delivered to the virtual earth of the operational amplifier, A, in Fig. 8.2.

If the time constant $C_F R_F$, in Fig. 8.2, is made very much greater than T_s, the output of the circuit will take up a voltage level $-\bar{i}R_F$. The circuit shown in Fig. 8.2 may now be looked at in two different ways.

If the switching frequency, $f_s = 1/T_s$, is taken to be the input variable to the circuit, and if v_s is taken to be a constant, V_s, then Fig. 8.2 may be thought of as a precision frequency to voltage converter. This follows because the output voltage will be given, using equation (8.5), by

$$v_o = -(C_s R_F V_s)f_s. \qquad (8.6)$$

This kind of voltage to frequency converter will be illustrated by the first experimental circuit of this chapter.

The second way of looking at the basic circuit shown in Fig. 8.2 is to consider the case of constant switching frequency, f_s, and then let the input signal voltage, v_s, be the variable. Always assuming that $C_F R_F \gg T_s$, the output from the circuit shown in Fig. 8.2 may then be written

$$v_o = -(R_2/R_1)v_s \qquad (8.7)$$

where $R_2 = R_F$ and $R_1 = 1/C_s f_s$. From this point of view, the switched capacitor arrangement at the input of the operational amplifier, shown in Fig. 8.2, is behaving as though it were a 'resistor'

$$R_1 = 1/C_s f_s. \qquad (8.8)$$

Equation (8.8) is a very important result because a 'resistor' which has a value determined by a capacitance, C_s, and a frequency, f_s, can be made

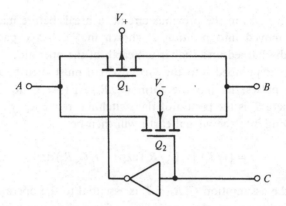

Fig. 8.3. *A CMOS bilateral switch.*

very precise and will be linear and temperature independent [5, 6]. What is even more exciting is that the value of this 'resistor' can be varied over a wide range by simply varying the switching frequency, f_s. This idea leads the circuit designer to begin proposing all kinds of switched capacitor versions of well-known analog circuits, like active filters, oscillators, phase locked loops (PLLs), and so on, which, in their conventional form, would have their performance controlled by variable resistors.

It is circuits of this kind that occupy the main area of interest in the field of switched capacitor circuits, as can be seen from the first six chapters of the text by Allen and Sánchez-Sinencio [4]. It should be mentioned that the literature, and also the background work in this field, is quite overwhelming. A very interesting review of this has been published by Weinrichter [7], who pointed out that the concept expressed here by equation (8.8) can be found in J. C. Maxwell's classical work, *Electricity and Magnetism*, which was first published in 1873.

8.4 Switch realisation

A single FET may be used as a switch, and this is the solution adopted in many switched capacitor circuits. Two problems do come up here, however. The first is the unavoidable charge injection from the gate of the transistor into the signal line which is being switched. The second problem is that performance must be influenced by the voltage level on the signal line, because this must change the gate to source voltage.

A way of getting around these two problems, while at the expense of more circuit complexity and increased parasitic capacitance, is to use a pair of CMOS transistors, as shown in Fig. 8.3. Here the p-channel enhancement MOST, Q_1, has its n-type substrate connected to V_+, and its gate must be made negative in order to turn it fully on. The opposite applies to Q_2, the n-channel enhancement MOST. This has its p-type substrate taken to V_-, and its gate must be made positive in order to turn it fully on. The two gates can then be driven, as shown in Fig. 8.3, by simply separating them with a standard CMOS inverter. The control input, C, then accepts the standard CMOS input levels [8].

Such a CMOS bilateral switch is quite symmetric. It should be possible to balance the charge injected into the signal line from the gate of Q_2 with the charge removed by the gate of Q_1. Similarly, the voltage level on the signal line, A to B, should not affect the characteristics of the switch because changes in the gate to source voltage at Q_1 should be compensated by equal and opposite changes at Q_2.

Two switches, of the kind shown in Fig. 8.3, would be needed to realise one change-over switch. It would also be necessary to arrange the control signals to these two switches in such a way that the essential break before make property is achieved. One of the many integrated circuits that are available to solve all these problems for the experimentalist is the LTC1043 [9]. This device contains two pairs of change-over switches, together with all the drive circuits needed, and an internal oscillator. Most applications, however, call for an external drive, and the LTC1043 can accept input frequencies up to a few MHz.

Armed with this valuable device, which also has an excellent application note to go with it [10], it is possible to look at a number of experimental circuits in the switched capacitor field. The first of these is a practical realisation of the basic charge pumping circuit, first shown in Fig. 8.2, when this is used as a frequency to voltage converter.

8.5 An experimental frequency to voltage converter

Fig. 8.4 shows the first experimental circuit for this chapter. The LTC1043 has a ± 4.7 V power supply to pins 4 and 17, set up by means of the two Zener diodes which are powered from the ± 15 V supply to the whole circuit. The OP07 is powered directly from the ± 15 V lines, to pins 7 and 4, not shown in Fig. 8.4, and these pins should be decoupled to ground in the usual way.

The clock input to the LTC1043 is taken to pin 16 through an 8.2 kΩ resistor. This makes it possible to drive the circuit from a simple laboratory square wave generator. The 8.2 kΩ resistor simply protects the LTC1043 against overdrive.

Only one half of the LTC1043 is used in this experimental circuit. In one position of the pair of switches, C_2 is charged up with the top of C_s

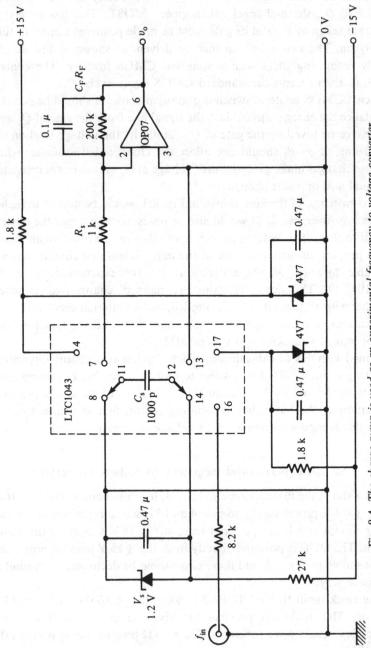

Fig. 8.4. The charge pump is used as an experimental frequency to voltage converter.

grounded and the bottom taken to -1.2 V. This -1.2 V comes from a voltage reference of the so-called bandgap variety [11]. In the second switch position, C_s is discharged into the virtual earth of the OP07, through R_s, but now the bottom of C_s is grounded and the top of C_s is connected to R_s. This illustrates the way in which two change-over switches may be used to invert a voltage in switched capacitor technique. The output from the experimental circuit, v_0, will be negative, although V_s is negative and the OP07 is being used as an inverting amplifier.

Note that pin 13 on the LTC1043 has been chosen as a grounded pin because this will disenable the charge balancing circuitry provided in the device. This is not needed in this simple application.

The values shown in Fig. 8.4 make $C_F R_F = 20$ ms. This allows the circuit to give a reasonably smooth output voltage when f_{in} is taken down as low as 500 Hz. Using equation (8.6), and substituting values for C_s, V_s and R_F, it is clear that the conversion constant of the experimental frequency to voltage converter should be 0.24 V/kHz. This means that the OP07 output voltage will limit when f_{in} is greater than 50 kHz. The linearity of the experimental circuit should be checked over this frequency range: 500 Hz–50 kHz.

So far the experimental circuit has been considered in the steady state: f_{in} is constant and v_0 is constant. Even when this is true, the circuit's dynamics must be considered because of the sampled data nature which is intrinsic to any switched capacitor circuit. Considerations along these lines bring up the reason for including R_s in the circuit. Surely the circuit would work better with $R_s = 0$ so that C_s could discharge as rapidly as possible?

Looking at pin 2 of the OP07 with a sensitive wide-band oscilloscope will show the function of R_s clearly. If R_s is short circuited, the virtual earth of the OP07 will be seen to rise by some 600 mV, as C_s discharges, and this considerable overload of the OP07 input will be seen to persist for some 3 μs. When R_s is made 1 kΩ again, the input overload is completely over in 1.5 μs. Viewing pin 7 on the LTC1043 will show a smooth exponential decay from about 700 mV with a time constant of 1.5 μs. This shows that, with C_s at 1000 pF, the total resistance in series with C_s, as it discharges into the virtual earth, is around 1.5 kΩ. With $R_s = 1$ kΩ, this implies that each switch contributes 250 Ω. This agrees with the data given for the LTC1043.

The experimentalist should also look at the effect of replacing the OP07 with a faster precision operational amplifier, like the OP15, LF356 or OP27. The input overload is almost unobservable with these fast amplifiers, always provided that $R_s = 1$ kΩ is still in circuit.

The transfer function, $F(s)$, of the frequency to voltage converter shown in Fig. 8.4 may be obtained by considering the currents flowing into the virtual earth of the OP07. These must sum to zero, so that

$$C_s V_s f_{in} + v_o/R_F + C_F \, dv_o/dt = 0. \tag{8.9}$$

Writing $\tilde{f}_{in}(s)$ as the Laplace transform of the varying input frequency, $f_{in}(t)$, and $\tilde{v}_o(s)$ as the Laplace transform of the output voltage, $v_o(t)$, it follows from equation (8.9) that

$$F(s) = \tilde{v}_o/\tilde{f}_{in} = -C_s R_F V_s/(1 + sC_F R_F) \tag{8.10}$$

and this transfer function will be used later in this chapter when the circuit shown in Fig. 8.4 is used again as part of another experimental circuit.

Finally, the ripple on the output of the experimental circuit shown in Fig. 8.4 should be looked at carefully. Provided $f_{in} \gg 1/C_F R_F$, this ripple should be a saw-tooth waveform, and should have *constant* peak to peak amplitude of about 12 mV, for the values of C_s, V_s and C_F given in Fig. 8.4, regardless of the value of f_{in}. It is easy to show that the peak to peak amplitude of the ripple is given by $V_s(C_s/C_F)$. It is independent of the value of R_F. Increasing R_F will, of course, increase the mean output level of the circuit shown in Fig. 8.4, for a given input frequency, and so decrease the percentage ripple on the output. Such an increase in R_F reduces the maximum input frequency that the circuit can handle. The best way to increase the useful input frequency range is to increase C_F, but this slows down the response of the circuit to changes in f_{in}. Considerations of this kind come up in the next section.

8.6 An experimental voltage to frequency converter

The frequency to voltage converter that was put forward as an experimental circuit in the previous section had remarkable linearity. The conversion constant, under steady state conditions, given by equation (8.10) as $-C_s R_F V_s$ V/Hz, depends upon three well-defined circuit values, and really linear behaviour should be observed for input frequencies well above $1/2\pi C_F R_F$, right up to the input frequency which causes limiting at the OP07 output.

This linear behaviour is in great contrast to the non-linear characteristic of an earlier experimental circuit described in this book: the triangle waveform generator of chapter 7 which had the voltage to frequency characteristic shown in Fig. 7.4. This observation brings up the possibility of linearising the non-linearity of a voltage to frequency converter, by connecting a very linear frequency to voltage converter across it as a feedback network.

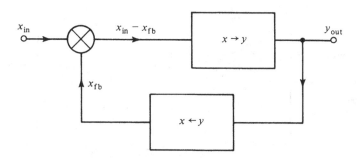

Fig. 8.5. *The general feedback system where input, x, and output, y, may have different dimensions.*

It should be mentioned at this point that the above idea is just one aspect of a very general idea. This is illustrated in Fig. 8.5 by the general feedback system shown there. In Fig. 8.5 an input variable, x_{in}, which could be a voltage, pressure, frequency, or any physical variable, is transformed into an output variable, y_{out}. The dimensions of x and y may be the same, as they are in the familiar feedback voltage amplifier problem, or they may be quite different: voltage and frequency, for example. The essential points in Fig. 8.5 are the feedback and the fact that the $x \to y$ function block has very high gain. The $y \to x$ feedback block is the part that determines the overall system behaviour. If the gain of the $x \to y$ function block is very high, $x_{in} - x_{fb}$ must be negligible so that x_{fb} must equal x_{in}. It then follows that y_{out} is forced to follow x_{in} with the same relationship that the feedback block, $y \to x$, sets between its input and output. The feedback system shown in Fig. 8.5 has a transfer function which is the *inverse* of the transfer function of its feedback network.

The difficulty of applying this important general principle in practice is the stability of the feedback system. This is a familiar fact to all workers in electronic circuits because of the problem of stability in connection with operational amplifiers, which are always used with feedback. In the operational amplifier case, x and y are both voltages and the feedback is done with a simple linear potential divider.

Fig. 8.6 shows how the feedback idea can be applied in the case of voltage to frequency conversion. Comparing Figs. 8.5 and 8.6, the input variables, x_{in} and v_{in}, are now both voltages and the output variables, y_{out} and f_{out}, both frequencies. The very high gain, and the differential function, both called for by the path x_{in} to y_{out} in Fig. 8.5, are realised in Fig. 8.6 by means of a simple integrator: the OP07 operational amplifier

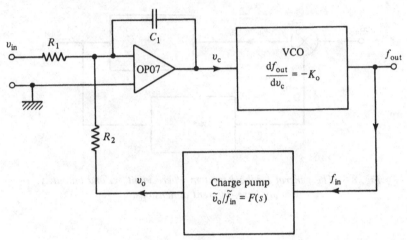

Fig. 8.6. *An experimental ultra-linear VCO can be built as a control loop using the VCO of Fig. 7.2, the charge pump of Fig. 8.5 and one simple integrator.*

with its feedback components, R_1, R_2 and C_1. At the virtual earth of the OP07,

$$v_{in}/R_1 + v_o/R_2 + C_1\,dv_c/dt = 0 \tag{8.11}$$

must hold.

The output frequency, in Fig. 8.6, is related to v_c by the non-linear relationship shown earlier in Fig. 7.4. In the steady state it is clear that this non-linearity will vanish from the problem because v_c is constant and equation (8.11) is simply

$$v_o = -v_{in}(R_2/R_1). \tag{8.12}$$

Now, again in the steady state, equation (8.10) reduces to

$$f_{in} = -v_o/C_s R_F V_s \tag{8.13}$$

and, of course, in Fig. 8.6 the f_{in} of the charge pump circuit, previously Fig. 8.4, is the f_{out} of the VCO, previously Fig. 7.2. Combining equations (8.12) and (8.13) thus gives

$$f_{out} = v_{in} R_2/C_s R_1 R_F V_s \tag{8.14}$$

as the characteristic of the ultra-linear VCO which should be realised by this new experimental circuit, when it is working under steady state conditions.

8.7 Stability problems

Some more theoretical work must be done before the experimental circuit shown in Fig. 8.6 can be built. This concerns its stability and is not simple because Fig. 8.6 shows a non-linear control system: it involves the VCO which has the non-linear voltage to frequency characteristic shown in Fig. 7.4. This characteristic has a slope, using the symbolism of Fig. 8.6,

$$df_{out}/dv_c = -K_o \tag{8.15}$$

where K_o is very nearly constant, at about 5 kHz/V, over the range 2–20 kHz. At least the *change* in output frequency, for this VCO, depends fairly linearly upon a *change* in input control voltage, v_c. It follows that equation (8.15) may be used to write

$$\tilde{v}_c = -\tilde{f}_{out}/K_o. \tag{8.16}$$

Equations (8.11), (8.10) and (8.16), again writing $f_{in} = f_{out}$, then combine to give

$$\frac{\tilde{f}_{out}}{\tilde{v}_{in}} = \frac{R_2}{C_s R_1 R_F V_s} \left(\frac{1 + sT_F}{1 + sT_D + s^2/\omega_N^2} \right) \tag{8.17}$$

as the transfer function of the system shown in Fig. 8.6, where the filter time constant

$$T_F = C_F R_F \tag{8.18}$$

the damping time constant,

$$T_D = C_1 R_2 / C_s K_o R_F V_s \tag{8.19}$$

and the natural resonant frequency,

$$\omega_N^2 = (C_s K_o V_s)/(C_1 C_F R_2) \tag{8.20}$$

all need to be chosen to optimise the system's dynamic performance. A well-designed system will show a rapid response to changes in input control voltage, with only a small overshoot. The response should not be oscillatory on the one hand nor overdamped on the other.

8.8 Measurements on the voltage to frequency converter

The value of $C_F R_F$ that was used in the first experimental circuit of this chapter, Fig. 8.4, was 20 ms. Now that Fig. 8.4 is part of the second experimental circuit, Fig. 8.6, it is sensible to leave $C_F R_F$ at 20 ms and make the system's damping time constant, T_D, also about 20 ms. With the values that have been chosen already ($C_s = 1000$ pF, $K_o = 5$ kHz/V, $R_F = 200$ kΩ and $V_s = 1.2$ V) the value of $C_s K_o R_F V_s$, in equation (8.19), is 1.2, and is dimensionless. So, if T_D is to be about 20 ms, a sensible choice for $C_1 R_2$, in equation (8.19) will also be 20 ms: for example, $C_1 = 0.1$ μF and $R_2 = 200$ kΩ. R_1 is made equal to R_2 so that $v_{in} = -v_o$ in the steady state. Note that v_{in} will thus be positive in this experiment.

This choice of values puts the natural resonant frequency of the system (equation (8.20)) close to 9 Hz, and the system should be stable.

Using the values given above, the first thing that should be measured on this new experimental circuit is its steady-state performance as a voltage controlled oscillator. In contrast to the non-linear voltage–frequency characteristic shown in Fig. 7.4, for the VCO of Fig. 7.2, the new system should have a very linear characteristic over the range 500 Hz–20 kHz. The upper frequency limit is the limit of the original VCO, as shown by Fig. 7.4, while the lower limit is where the ripple on the output of the charge pump, discussed at the end of section 8.5, begins to be significant. The voltage–frequency characteristic of this new system should extrapolate back to pass exactly through the origin, again in contrast to Fig. 7.4.

The low frequency limit can be reduced by increasing the time constants T_F and T_D, with an unavoidable reduction in system bandwidth. The bandwidth of this system is, of course, its ability to follow changes in v_{in} and reflect these changes accurately in the way that f_{out} changes. This aspect of system performance is easily measured by using a square wave generator, which has an offset facility, as an input signal. The component values being used at present make the slope of the voltage–frequency characteristic, $R_2/C_s R_1 R_F V_s$, equal to 4.17 kHz/V. An offset of $+2$ V on the square wave generator source will thus take f_{out} close to the centre of its range. A low level square wave on top of this $+2$ V, a few hundred millivolts in amplitude, for example, will then allow the dynamic system response to be seen if v_c, in Fig. 8.6, is viewed with an oscilloscope. The frequency of the square wave input must be low: about 1 Hz. A damped response should be observed, with a small overshoot, and with a response time of about 100 ms.

Another way of observing the dynamic performance is to observe the output from the system, f_{out}, when v_{in} is a low amplitude square wave at 1 Hz, added to an offset of a few volts. The oscilloscope should now be run on a much faster timebase, so that the real change in f_{out} can be seen. The output waveform will be seen to change frequency in quite a complex way, each time v_{in} changes level. A more subjective way of appreciating this complex behaviour is actually to listen to the output of the system. This is easily done by connecting a low impedance loudspeaker to the output through a 2 kΩ resistor.

The main conclusion to be drawn is that the precision voltage to frequency performance which has been obtained here has been bought at the cost of speed of response. The VCO which has been used at the heart of this experimental system was the first experimental circuit of chapter 7:

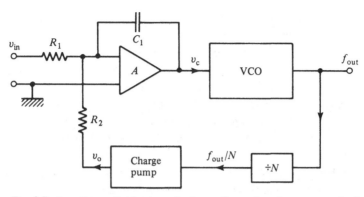

Fig. 8.7. Inserting a divider into the loop allows ultra-linear control of very high frequency voltage controlled oscillators.

the voltage controlled triangle waveform generator of Fig. 7.2. This waveform generator could change frequency very rapidly when its input control voltage was changed. In contrast, the precision VCO of the present chapter is a closed loop control system and its dynamic performance is dominated by the high gain in its input circuit, the OP07 which is being used as an integrator, and then again by the precision frequency to voltage converter which is used as the feedback path. This converter is a switched capacitor circuit which, by its very nature as a sampled data system, must include a low pass filter: the pass band must be well below the sampling frequency.

One final point concerning control loops which make use of frequency to voltage converters is the possibility of inserting a frequency divider into the loop. This is illustrated in Fig. 8.7. The input frequency to the charge pump is now only a fraction of the VCO frequency, so that the VCO can now operate at a very high frequency, this being the system output frequency, while the charge pump circuit continues to operate in the kilohertz region. Williams [12] has shown how this idea can be used to make an ultra-linear voltage to frequency converter, operating over the 100 kHz–1.1 MHz range, with a linearity of 7 ppm, despite the fact that the VCO at the heart of the system is a crude relaxation oscillator.

8.9 Switched capacitor filters

The possibility of making electronic filters by means of the switched capacitor technique was probably first put forward by Fried [13] in 1972. A review article by Solomon [14] claims that switched capacitor is now the first choice technology for filter realisation, because of the cheapness, compactness and precision that may be obtained.

The method may be introduced by considering the simple second-order active filter shown in Fig. 8.8. This is arranged as a bandpass filter and involves three operational amplifiers with feedback networks using only resistors and capacitors [15]. The labelling of the components in Fig. 8.8 may look slightly odd. The reason is that this is the circuit which is going to be used as the prototype for the last experimental circuit of this chapter, and this will involve only two operational amplifiers, A_1 and A_2, and only five capacitors. The components labelled A and R in Fig. 8.8 will be eliminated.

From the simplest point of view, a switched capacitor filter is developed from an active filter, of the kind shown in Fig. 8.8, by the technique of replacing all the resistors, which are connected to the virtual earths of the operational amplifiers, by switched capacitors. This idea was discussed in section 8.3, after equation (8.8) had been obtained, this equation showing that a resistor, R, could be replaced by a capacitor, C_s, and a switch, operating at a frequency f_s, because of the identity between R and $1/C_s f_s$.

Fig. 8.9 illustrates this idea in a particular application. The integrator on the far right of Fig. 8.8 is now shown alongside its switched capacitor equivalent. The simple integrator in Fig. 8.9(a) has the transfer function, to an input signal $v_1(t) = \hat{v}_1 \exp(j\omega t)$

$$\hat{v}_2/\hat{v}_1 = -1/j\omega C_5 R_4. \qquad (8.21)$$

In Fig. 8.9(b), the resistor R_4 is replaced by capacitor C_4 and the switch, S, which is a break before make switch changing position repetitively at frequency f_s. According to equation (8.8), this is identical to a resistor $1/C_s f_s$ so that, provided $f_s \gg \omega/2\pi$, the circuit shown in Fig. 8.9(b) will have a transfer function

$$\hat{v}_2/\hat{v}_1 = -C_4 f_s/j\omega C_5. \qquad (8.22)$$

Note that the inverting property, the minus sign in equations (8.21) and (8.22), is maintained. This is an important point because a closed loop active filter, like the one shown in Fig. 8.8, must always have an odd number of inverting operational amplifiers in its loop to ensure negative feedback. The first amplifier, A, in Fig. 8.8 is, in fact, doing nothing but inversion.

In switched capacitor technique, inversion may be done without an extra amplifier. This has already been illustrated in the first experimental circuit of this chapter, Fig. 8.4. Fig. 8.10 illustrates the point for the particular problem which is under consideration now: the switched capacitor realisation of the filter shown in Fig. 8.8. A *non-inverting*

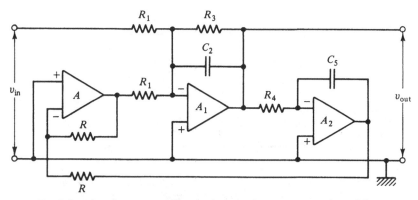

Fig. 8.8. *A bandpass active filter built using three operational amplifiers with RC feedback networks.*

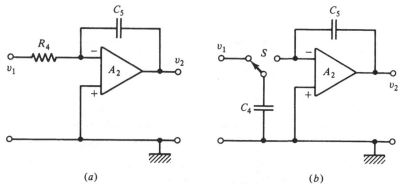

(a)   (b)

Fig. 8.9. *The simple integrator (a) may be replaced by the switched capacitor circuit (b) provided the switching frequency is very much greater than the highest frequency component in the input, $v_1(t)$.*

integrator may be realised if *two* switches are used in the switched capacitor version, instead of the single switch shown in Fig. 8.9(b). The circuit shown in Fig. 8.10 will have a transfer function

$$\hat{v}_2/\hat{v}_1 = +C_1 f_s/j\omega C_2 \qquad (8.23)$$

provided $f_s \gg \omega/2\pi$. An excellent review of all these possibilities, together with the exact theory for finite switching frequency, has been given by Martin [16]. More recent work has been reviewed briefly in an important paper by Psychalinos and Haritantis [17].

Finally, the active filter circuit shown in Fig. 8.8 includes the resistor R_3 which controls the damping in this filter, and thus determines its Q. The way in which this resistor, R_3, may be replaced by a switched capacitor, C_3, is shown in Fig. 8.11.

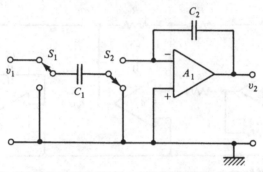

Fig. 8.10. *A non-inverting integrator in switched capacitor technique involves two break before make switches.*

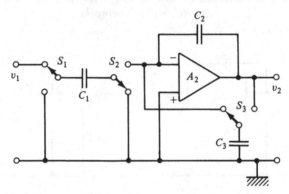

Fig. 8.11. *Damping is added to the non-inverting integrator by means of a third capacitor, C_3, and its switch, S_3.*

8.10 Filter theory

No mention has been made so far of the theory behind the active filter shown in Fig. 8.8. How are the component values for this bandpass filter to be determined? This, as throughout this book, is in the spirit of the 'circuit shape' ideas put forward in section 1.3. When a new idea in electronic circuit design comes up, it is being thought about as a sketch, a circuit shape, or circuit idea. The detailed calculation of how this idea works can only be tackled once this first stage, of sketching and arranging things, is done. This is what has been happening here in Figs. 8.9, 8.10 and 8.11.

Going back now to the simple active filter shown in Fig. 8.8, the overall transfer function of this filter is easily found by summing all the currents into the virtual earth of A_1 to zero. Assuming that the input is $\hat{v}_{in} \exp(j\omega t)$ and the output is $\hat{v}_{out} \exp(j\omega t)$, this summing may be expressed as

$$\hat{v}_{in}/R_1 + \hat{v}_{out}/j\omega C_5 R_4 R_1 + j\omega C_2 \hat{v}_{out} + \hat{v}_{out}/R_3 = 0. \qquad (8.24)$$

Equation (8.24) is considerably simplified if it is assumed that $C_2 = C_5$ and $R_1 = R_4$, in Fig. 8.8. This would be the usual choice. The natural resonant frequency of the filter is then

$$\omega_N = 1/C_2 R_1 \tag{8.25}$$

and the Q is

$$Q = R_3/R_1. \tag{8.26}$$

These expressions for ω_N and Q may then be put into equation (8.24) to give the transfer function for the filter as

$$\frac{\hat{v}_{out}}{\hat{v}_{in}} = \frac{-j\omega/\omega_N}{(1-\omega^2/\omega_N^2)+j(\omega/\omega_N Q)}. \tag{8.27}$$

The filter thus has an inverting gain of Q at its centre frequency, $\omega = \omega_N$. Its bandwidth, in hertz, is $\omega_N/2\pi Q$.

8.11 The switched capacitor version

After this theoretical digression, the method of working can go back to circuit shapes and the circuit ideas expressed in Figs. 8.9, 8.10 and 8.11. These can all be used to convert Fig. 8.8 into the switched capacitor version of one and the same bandpass filter. The result is shown in Fig. 8.12, which is the final experimental circuit for this chapter.

Fig. 8.12 shows a switched capacitor filter using two CA3140 operational amplifiers. These have the MOS input transistors which are essential for switched capacitor applications. A single LTC1043 device is used to implement the four switches which are needed. It will be clear that Fig. 8.12 is the result of combining Figs. 8.9, 8.10 and 8.11, working progressively from Fig. 8.8 to Fig. 8.12. The resistors R_1, R_3 and R_4, in Fig. 8.8, have been replaced by capacitors C_1, C_3 and C_4 in Fig. 8.12. The first amplifier, A, in Fig. 8.8 and its two resistors, both labelled R, have been replaced by the two switch arrangement, first shown in Fig. 8.10.

The natural resonant frequency of this switched capacitor filter will be given by

$$\omega_N = f_s C_1/C_2 \tag{8.28}$$

which follows from equation (8.25) when R_1 is replaced by its switched capacitor equivalent, $1/f_s C_1$. Equation (8.28) makes it immediately clear how easily a switched capacitor filter may be tuned. The switching frequency is now the variable which controls ω_N. The ratio C_1/C_2 will be constant and must, of course, be kept small so that $\omega_N/2\pi$, the frequency at which the filter is used, is always small compared to f_s. There will be

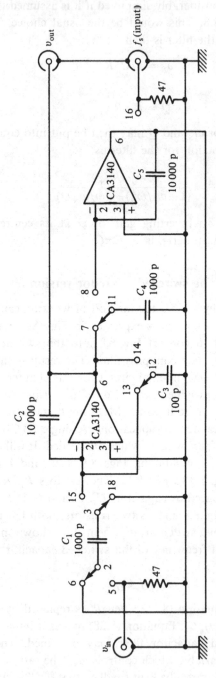

Fig. 8.12. An experimental switched capacitor bandpass filter using two CA3140 operational amplifiers and one LTC1043. The LTC1043 is powered by a ±4.7 V supply, as in Fig. 8.4, and the CA3140s are powered from ±15 V. All devices have the usual decoupling at their power supply pins. Only the signal pins of the devices are shown here, with their numbers.

much to say on this point when experimental work is done with the circuit shown in Fig. 8.12.

The Q of the switched capacitor filter shown in Fig. 8.12 will be given by

$$Q = C_1/C_3 \qquad (8.29)$$

when the relationships $R_3 = 1/f_s C_3$ and $R_1 = 1/f_s C_1$ are substituted into equation (8.26). This shows that the Q of this switched capacitor filter does not depend upon the switching frequency, f_s. The filter may be tuned, by varying f_s, but it should always have the same Q. This constant Q property is much more difficult to obtain with the so-called biquad RC active filter, shown in Fig. 8.8, because R_3, R_1 and R_4 all need to be varied together. Usually this RC filter is tuned by changing only R_1 and R_4. This gives the filter constant bandwidth as it is tuned [15].

8.12 Construction and experimental work

The layout of the experimental circuit shown in Fig. 8.12 should be given some thought because the wiring between devices is quite complicated. Note that pin 18 on the LTC1043 has been chosen as a grounded pin, and pin 13 is taken directly to a virtual earth. This choice should disenable the charge balancing circuitry in the LTC1043 which is not needed in this application.

The level of the switching input must match the input requirements of the LTC1043. These depend upon the power supply to pins 4 and 17. When these are at ± 4.7 V, a square wave input of 5 V amplitude, that is ± 2.5 V, is ideal.

When f_s is set at 100 kHz, the values shown in Fig. 8.12 should give a filter with a sharp bandpass response at a centre frequency of 1.59 kHz. This follows from equation (8.28). The gain at this centre frequency would be expected to be 10, from equation (8.29), because C_3 is 100 pF while C_1 and C_4 are both 1000 pF. It will be found, however, that the gain is much higher than this, and a clue as to why this is so can be obtained if C_3 is increased.

Making C_3 larger, say 200 pF and then 400 pF, will reduce the gain at ω_N, and increase the bandwidth, as would be expected, but the actual value of this gain will be found to match the theoretical expectations of equation (8.29) more and more as C_3 is increased. There is clearly some intrinsic positive damping in this switched capacitor filter. In fact, making C_3 only 50 pF will make the filter unstable.

The origin of this positive damping is in the sampled data nature of the system itself. Each capacitor being switched, first accepts its charge in one switch position, and then passes this charge on in the next half cycle of the switching input signal, f_s. This means, in the simplest case, that a time delay, $T_s/2$, is introduced by a switched capacitor input circuit. In most cases the time $T_s/2$ is quite negligible on the time scale of the signal frequency: the whole point is to use a switching frequency very much greater than the signal frequency. With a bandpass filter of the kind being considered here, however, the small time delay, $T_s/2$, represents an additional phase *lag* in the feedback loop. This lag is comparable with the small phase *advance* which C_3 introduces into the loop with the intention of obtaining high gain and narrow bandwidth at the bandpass frequency. These problems should be understood, and the first chapter of the text by Allen and Sánchez-Sinencio [4] will be valuable in this connection, as will be the review paper by Broderson, Gray and Hodges [18].

Despite the fact that the gain of the filter shown in Fig. 8.12 will be higher than expected, its performance will be found quite impressive. The centre frequency can be changed easily over the range 500 Hz–5 kHz by changing f_s. The gain at ω_N, and the filter Q, will be seen to be constant, over this entire range of tuning. The noise level at the output of the filter should be checked, when the input is short circuited. If the layout has been done well, the only noise of any importance at the output of the filter will be at f_s, and this should be only a few millivolts.

Finally, there is the interesting behaviour of the filter when an input signal close to the switching frequency, f_s, is deliberately introduced. If a signal of about 10 mV peak to peak is applied to the input of the circuit shown in Fig. 8.12, and the frequency of the input is increased, well above the value of ω_N and then still further until it equals and then exceeds f_s, a very remarkable observation will be made. As the input signal frequency passes through f_s, the output of the circuit shown in Fig. 8.12 will be a sine wave, the amplitude of this sine wave will be Q times the amplitude of the input, where Q is the gain of the filter previously measured at ω_N, but the *frequency* of this sine wave will be equal to $\omega_N/2\pi$: the bandpass frequency of the filter. This phenomenon is called *aliasing*: the process of sampling the input data, when the input data is changing at a frequency close to the sampling frequency, can reconstruct false signals at lower frequencies. A close examination of this effect, with the circuit shown in Fig. 8.12, will show that there are two input frequencies close to f_s at which an alias appears. One is $\omega_N/2\pi$ below f_s while the other is $\omega_N/2\pi$ above. This problem of aliasing is central to the field of sampled data systems: Horowitz and Hill give a brief but useful discussion in their book [19]. A detailed discussion may be found in the text by Allen and Sánchez-Sinencio [4].

8.13 Conclusions

This chapter has been dealing with just a few aspects of a new and exciting area in the field of electronic circuit design. Perhaps the most interesting feature of switched capacitor circuits is the fact that the most well-established principles, fundamental to electronic engineering, are involved. There is nothing in this chapter which Guglielmo Marconi would not have understood; in fact he would have been quite familiar with mechanically switched sampled data systems, operating up to 50 kHz, because these were used by one of his most important competitors, Rudolf Goldschmidt, before the First World War [20].

Marconi saw the future of electronics in the development of useful amplifying devices, working with continuous signals, and in this he was most certainly correct. It is interesting to reflect just how long this development has taken. Only since the mid-70s, has the electronic circuit designer been able to obtain an operational amplifier, with truly remarkable performance, as a small integrated circuit costing little more than the socket which might be used to mount it. Only in the last few years has it become possible to manufacture dozens of such high quality amplifiers, along with hundreds of other components, all together within a few square millimetres of silicon, and so realise switched capacitor filters of the advanced kind illustrated in Solomon's review paper [14].

But this is by no means the end of the story. Today, developments taking place in gallium arsenide integrated circuit processing have taken switched capacitor filter technology well up into the radio frequencies [21]. With gallium arsenide circuits, switching frequencies of several hundred megahertz are already being used, and it should be possible to enter the gigahertz area in the near future.

Notes

1 Gray, P. R., and Meyer, R. G., *Analysis and Design of Analog Integrated Circuits*, John Wiley, New York, second edition, 1984, pp. 737–64.
2 Soárez, R. E., Gray, P. R., and Hodges, D. A., *IEEE J. Sol. St. Circ.*, **SC-10**, 379–85, 1975.
3 Barbour, C. W., *Instruments and Control Systems*, **35**, No. 8, 104–5, August 1962.
4 Allen, P. E., and Sánchez-Sinencio, E., *Switched Capacitor Circuits*, Van Nostrand, New York, 1984.
5 Caves, J. T., Copeland, M. A., Rahim, C. F. and Rosenbaum, S. D., *IEEE J. Sol. St. Circ.*, **SC-12**, 592–9, 1977.
6 Hosticka, B. J., Broderson, R. W., and Gray, P. R., *IEEE J. Sol. St. Circ.*, **SC-12**, 600–8, 1977.

7 Weinrichter, H., *IEEE Circ. Syst. Mag.*, **CAS-M-6**, No. 4, 3–8, December 1984.

8 Jones, M. H., *A Practical Introduction to Electronic Circuits*, Cambridge University Press, Cambridge, second edition, 1985, pp. 227–8.

9 *1986 Linear Databook*, Linear Technology Corp., Milpitas, California, pp. 8.3–8.18.

10 Williams, J. M., Applications for a switched capacitor instrumentation building block, in: *Linear Applications Handbook*, Linear Technology Corp., Milpitas, California, 1987, pp. 3.1–3.16.

11 Easily available examples of 1.2 V bandgap voltage references devices are the Teledyne TSCO4BJ, The GE ICL8069DCZR, the Micro Power Systems MP5010GN and the Plessey ZN423T.

12 Williams, J. M., Designs for high performance voltage-to-frequency converters, in: *Linear Applications Handbook*, Linear Technology Corp., Milpitas, California, 1987, pp. 14.1–14.20. The circuit referred to is Fig. 10 of this Application Note.

13 Fried, D. L., *IEEE J. Sol. St. Circ.*, **SC-7**, 302–4, 1972.

14 Solomon, C. W., *IEEE Spectrum*, **25**, No. 6, 28–32, June 1988.

15 Horowitz, P., and Hill, W., *The Art of Electronics*, Cambridge University Press, Cambridge, second edition, 1989, pp. 276–9.

16 Martin, K., *IEEE Trans. Circ. Syst.*, **CAS-27**, 237–44, 1980.

17 Psychalinos, C., and Haritantis, I., *IEEE Trans. Circ. Syst.*, **CAS-36**, 1493–4, 1989.

18 Broderson, R. W., Gray, P. R., and Hodges, D. A., *Proc. IEEE*, **67**, 61–75, 1979.

19 Note 15 above, pp. 283 and 775–6.

20 Mayer, E. E., *Proc. IRE*, **2**, 69–108, 1914. The pages of interest are 85–92.

21 Haigh, D. G., Toumazou, C., and Betts, A. K., Switched capacitor circuits and operational amplifiers, in: D. Haigh and J. Everard (Eds.) *GaAs technology and its impact on circuits and systems*, Peter Peregrinus, London, 1989, pp. 313–56.

9

Phase locked loop circuits

9.1 Introduction

The phase locked loop (PLL) is used in a number of instrumentation systems. Best's text on PLLs [1] has a chapter on applications in which he lists tracking filters; modulators and demodulators for AM, FM and PM; the recovery of weak signals; frequency synthesis, a field on which there is another valuable text [2]; motor speed control; stereo decoding; sub-carrier detection in colour television; and many others.

The original idea of the PLL probably came from the work of de Bellescize in 1932 [3], who used the PLL in AM receiver design. Later work on this application by Tucker and Ridgway provoked a considerable correspondence [4] which suggests that the PLL may be a case of multiple invention, as is so often the case in the field of electronic circuit design.

The simplest kind of PLL is shown in Fig. 9.1. A phase detector is used to give a measure of the phase difference, φ, between an incoming signal, $v_1 \sin(\omega t)$, and a VCO. This phase difference is used as an error signal in a closed loop control system in such a way that the VCO 'locks' on the incoming signal and follows its changes in phase, and thus in frequency, over some range determined by the designer.

This chapter begins with a very brief look at PLL theory, and this leads at once to an experimental programme using an easily available PLL integrated circuit. This experimental programme illustrates a number of problems in PLL design and leads to the main topic of this chapter, which is a review of some of the circuit shapes, or circuit ideas, that have been applied in PLL integrated circuits over the past 20 years.

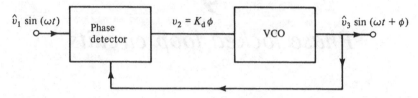

Fig. 9.1. *A first order PLL.*

9.2 The low pass filter

The PLL is usually introduced in the text books [5, 6] with a low pass filter
in between the phase detector and the VCO. It is important to bear in
mind why the need for this filter is tacitly assumed. The reason is that it
is impossible to build a phase detector which gives an instantaneous
measure of phase: that is one with the characteristic

$$v_2 = K_d \varphi \tag{9.1}$$

as shown in Fig. 9.1, where K_d is a simple constant, with dimensions volts
per radian, involving no time dependence. The best that may be done in
this direction is to build a digital phase detector that measures the time
between zero crossings of the signal input and the VCO, also measures the
period of one of these signals, and then gives a measure of phase,
according to some algorithm, as an analog output. Best deals with such
digital phase detectors in his book [7], but even such an advanced kind of
phase detector must introduce a time delay of at least one cycle between
the instantaneous values of its two inputs and the corresponding sampled
value of their phase difference.

A simple phase detector, for example, a combination of limiting
amplifiers and an analog multiplier, can only give a measure of relative
phase angle by means of averaging. To implement this averaging, a low
pass filter must be introduced into the PLL and this will have a profound
effect upon the loop dynamics.

The most straightforward way of dealing with the dynamics of the PLL
is to begin with the open-loop transfer function of the first order PLL
shown in Fig. 9.1. Fig. 9.2 shows the first order loop opened, with its input
and output variables now expressed as phases, φ_{in} and φ_{out}. For simplicity,
the feedback input is taken as the $\varphi = 0$ reference. The change in the VCO
frequency, shown in Fig. 9.2 as $\Delta f = K_o K_d \varphi_{in}$, where K_o is the control
constant of the VCO in hertz per volt, may be written in terms of phase
by using the fundamental relationship

$$2\pi\Delta f = d\varphi_{out}/dt. \tag{9.2}$$

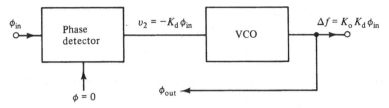

Fig. 9.2. *Opening the first order loop.*

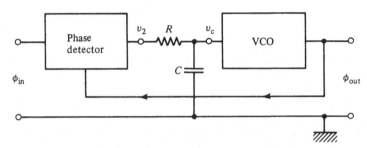

Fig. 9.3. *Adding a low pass filter to the PLL.*

By writing $\tilde{\varphi}_{in}$ and $\tilde{\varphi}_{out}$ to represent the Laplace transforms of the input and output variables, $\varphi_{in}(t)$ and $\varphi_{out}(t)$ the open-loop transfer function of the combined ideal phase detector, defined by equation (9.1), and VCO follows at once as

$$\tilde{\varphi}_{out}/\tilde{\varphi}_{in} = 2\pi K_o K_d/s \qquad (9.3)$$

where s is the complex variable of the Laplace transform.

Equation (9.3) shows that the VCO behaves like an *integrator*: its transfer function is of the form $1/s$.

9.3 Loop dynamics with a low pass filter

Fig. 9.3 shows the PLL with a low pass filter inserted into the loop: a simple RC network. This network has a transfer function

$$\tilde{v}_c/\tilde{v}_2 = 1/(1+sCR) \qquad (9.4)$$

where v_c is now the control voltage input to the VCO.

It follows that the open-loop transfer function of the PLL shown in Fig. 9.3 may be obtained by simply multiplying equation (9.3), which is the open-loop transfer function of the PLL with no filter, by equation (9.4). The result is

$$a(s) = 2\pi K_o K_d/s(1+sCR). \qquad (9.5)$$

Using the well-known [8] relationship between the closed-loop transfer function, $A(s)$, of a feedback system (with 100% negative feedback), and its open-loop transfer function, $a(s)$,

$$A(s) = a(s)/[1 + a(s)] \tag{9.6}$$

it follows that the closed-loop transfer function of the second order PLL, shown in Fig. 9.3, may be written

$$\tilde{\varphi}_{out}/\tilde{\varphi}_{in} = \omega_n^2/(\omega_n^2 + \alpha_1 \omega_n s + s^2) \tag{9.7}$$

where

$$\omega_n^2 = 2\pi K_o K_d/CR \tag{9.8}$$

$\omega_n/2\pi$ being the natural resonant frequency of this second order control loop, and

$$\alpha_1 = 1/\omega_n CR \tag{9.9}$$

giving a measure of the damping.

The designer would now consider what values of ω_n and α_1 would be optimum for the application under consideration. For example, ω_n might be chosen to correspond to the bandwidth expected for the incoming phase modulated signal, and then CR chosen to make $\alpha_1 = \sqrt{2}$, which is the value that gives a maximally flat frequency response. These system considerations may run into very severe limitations, however, because of the intrinsic properties of the circuits which make up the PLL: the VCO and the phase detector. It is the circuit detail of PLLs which is really the topic of this chapter. The theory which has been outlined above is only needed to guide the designer along broadly sensible lines. It turns out that PLLs are an excellent example of circuit considerations laying considerable constraints upon system performance.

In order to introduce this aspect of PLL circuit design, it is best to go at once to a very simple experimental circuit using one of the widely available integrated circuit PLLs.

9.4 The CD74HC4046A

There are a number of integrated circuit PLLs in which a VCO, one or more varieties of phase detector, and perhaps a voltage reference along with a few operational amplifiers, are all combined together in one monolithic circuit inside a single package. The CD74HC4046A is a fairly recent example [9], and is of particular interest because it has three quite different kinds of phase detector available. Because of this, the VCO control input is brought out to one of the 16 external pins. This is a particularly useful feature for the experimentalist because it makes it possible to insert a wide variety of filter types in between the phase detector and the VCO. Many integrated circuit PLLs do not have such flexibility because the phase detector output is connected internally to the VCO control input.

The CD74HC4046A is a CMOS integrated circuit. It is a development of the earlier CD4046A and CD4046B devices, which only have two kinds of phase detector available, and can only work up to just over 1 MHz. The CD74HC4046A can operate up to 18 MHz. Nearly all the experiments described below can be done with the earlier devices, however, using slightly different component values.

9.5 Experiments with the type I phase detector

Fig. 9.4 shows the first experimental circuit for this chapter. This uses the first kind of phase detector which is available on the CD74HC4046A device, shown on the left-hand side of Fig. 9.4, which involves two high gain amplifiers, A_1 and A_2, acting as limiting amplifiers, driving an exclusive OR gate, X. Marking this gate with an X indicates its close connection to the analog multiplier mentioned above in section 9.2.

The limiting amplifiers, A_1 and A_2, convert the two input signals into square waves. These square waves have the logic level of CMOS circuits: just above zero and just below V_{cc}. The exclusive OR gate gives a V_{cc} output if only *one* of its inputs is high. It thus produces a continuous zero output when the signals on the input pins, pins 14 and 3, are in-phase, and a continuous V_{cc} output when these two input signals differ in phase by 180°. In between these two extremes, the exclusive OR output will be a rectangular wave, at double the frequency of the input signals, varying in mark to space ratio as the phase varies, and having 1:1 mark to space ratio when the two input signals differ in phase by 90°.

It follows that this kind of phase detector, provided it is followed by a low pass filter which completely smooths its output voltage, will produce a *linearly* varying output voltage as the phase is varied. This is exactly the same as the ideal phase detector characterised by equation (9.1). The value of K_d will be V_{cc}/π. The output voltage will not vary about zero however, but about $V_{cc}/2$, which is, of course, the reference level for all the circuits in the CD74HC4046A. The input amplifiers, A_1 and A_2, for example, have their inputs referred to $V_{cc}/2$, which is why the 2200 pF coupling capacitor, shown in Fig. 9.4, is needed.

The first experimental measurement that should be made on the CD74HC4046A is to determine its VCO characteristic. To do this, the +5 V power supply should be applied to pins 16 and 8, decoupled at the

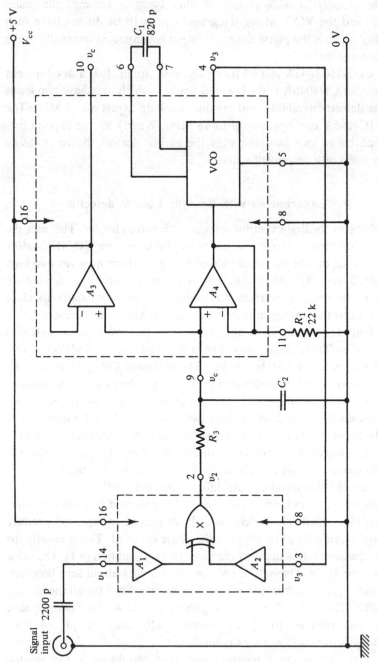

Fig. 9.4. The first experimental circuit using the CD74HC4046A. The VCO control voltage may be viewed on pin 10, which must be taken to ground through 10 kΩ.

device pins as usual, and pin 5 should be grounded to enable the VCO. The VCO is a relaxation oscillator which has a single timing capacitor, C_1, and its frequency is controlled by the current drawn from it by the output of A_4. The circuit details and waveforms of the VCO will be considered in section 9.13. The interest now is in the behaviour of this PLL as a system.

The choice of $C_1 = 820$ pF and $R_1 = 22$ kΩ, shown in Fig. 9.4, should make the free running VCO frequency close to 200 kHz. To observe this, omit the low pass filter, $R_3 C_2$ in Fig. 9.4, for the moment and take pin 9 to $+2.5$ V. A square wave of period close to 5 μs should be observed on pin 4, having a mark to space ratio very close to 1:1, and with a high level just below $+5$ V and a low level just above zero.

Now vary the voltage on pin 9 about the present level of 2.5 V. For small departures from $+2.5$ V, a fairly linear variation in frequency, with K_o close to 85 kHz/V, should be observed. When v_c approaches $+4$ V, or falls as low as $+1$ V, the characteristic becomes very non-linear, with K_o increasing considerably. The characteristic is shown in the data sheet [9] for various values of V_{cc}, and it is worthwhile examining this non-linear feature in some detail.

9.6 Inserting the low pass filter

The first experiments with the type I phase detector should be made with a simple RC low pass filter, of the kind shown in Fig. 9.4. The time constant, $C_2 R_3$, should be made very much greater than the free running period of the VCO. Such a choice makes it possible to assume that the control voltage, v_c, is effectively smoothed and is free from any ripple at the double frequency. The value $K_d = V_{cc}/\pi$ may then be attributed to the phase detector.

It will then follow that the natural resonant frequency of the PLL, given by equation (9.8), will be very low compared to the VCO frequency, and that the loop will be very lightly damped. For example, a choice of $R_3 = 47$ kΩ and $C_2 = 0.1$ μF will make $\omega_n/2\pi$ just over 2 kHz, or about 1% of the VCO centre frequency. The damping parameter, given by equation (9.9), will be less than 0.02. Nevertheless, closing the loop, to make up the complete experimental circuit shown in Fig. 9.4, will give a stable PLL and make possible some interesting measurements.

To begin with, set the input signal frequency to be at least 100 kHz higher, or lower, than the 200 kHz free running frequency of the VCO. View the VCO output, on pin 4, and the input signal with an oscilloscope. The timebase should be triggered from the input signal channel, Y_1. On first switching on the power supply, the display should show the input signal, which should be a sine wave set to 200 mV peak. The VCO should appear as two lines across the display, because it is running at a completely different frequency to the input signal: changing the timebase trigger to Y_2 will show that the VCO is running at its free running frequency.

Now shift the input signal frequency slowly towards the VCO free running frequency. When the two frequencies almost coincide, the VCO square wave will suddenly appear on the display with its positive and negative going edges almost coincident with the input signal positive and negative peaks. The VCO is lagging the input signal by close to 90° and the loop is now in lock.

Confirm that a double frequency square wave is now present at the phase detector output and that pin 9 is, consequently, at $+2.5$ V. It is good practice always to view what is happening on pin 9 by looking at pin 10, that is through the buffer amplifier A_3. In some of the experiments that follow, it is very important to maintain pin 9 as a very high impedance input.

9.7 Observation of the capture process

Once the PLL of Fig. 9.4 is in lock, it will be possible to vary the input frequency over quite a wide range, at least 100 kHz–300 kHz, and observe the input sine wave and the VCO square wave staying locked together in frequency. Only the relative phase of these two signals will vary.

To begin to see the detail of what is happening, the VCO control voltage at pin 10 should be measured as a function of signal input frequency. Ideally, this should be done with a swept frequency source and an oscilloscope. The result should look something like that shown in Fig. 9.5. Once in lock, v_c will vary linearly as the input frequency varies, limited between $+1$ V and $+4$ V. These limits are determined by the values of v_c at which the VCO characteristic goes very non-linear, as the first experimental measurements, described in the previous section, have shown. If the input signal frequency goes outside the lock range, shown in Fig. 9.5, the output of the low pass filter goes to $+2.5$ V. This happens because the output of the exclusive OR gate, which is the input to the low pass filter, is then a square wave at the beat frequency between the input signal, at say 50 kHz or 350 kHz, and the free running VCO at 200 kHz.

Entering the lock range from outside, v_c will be seen to remain at $+2.5$ V until the input frequency gets very close to the VCO free running frequency. A small jump then occurs: negative going if the approach is from below, positive going if from above. These two small jumps are shown in Fig. 9.5 as defining the 'capture range', which is, in fact, very small with the component values shown in Fig. 9.4 and with $R_3 = 47$ kΩ

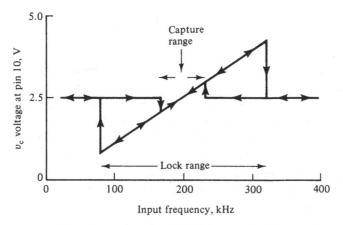

Fig. 9.5. *Typical measurements of VCO control voltage as the input signal frequency to the PLL is varied slowly.*

and $C_2 = 0.1\ \mu\text{F}$. For clarity, the capture range has been exaggerated in Fig. 9.5, and the figure also omits to show some small fluctuations, which will be seen on v_c, that indicate where the loop attempts to lock on input frequencies having some kind of harmonic relation to the free running VCO frequency. These harmonic relations may be very complicated and deserve close investigation.

The capture process is one of the most fascinating features of a PLL, and perhaps the most difficult to handle theoretically. Its complexity may be seen if v_c, the voltage at pin 10 in Fig. 9.4, is viewed with the oscilloscope on a timebase of about 1 ms/div, and on a sensitivity (a.c. coupled) of a few millivolts per division. Entering the lock range from outside, and approaching the VCO free running frequency, a very small sinusoidal signal will be observed, growing in amplitude and falling in frequency as the VCO free running frequency is approached. This is the beat frequency between the input signal and the VCO becoming low enough to pass through the low pass filter with less and less attenuation as its frequency falls.

Eventually, this fluctuation in v_c will be producing enough frequency modulation of the VCO to cause it, periodically, to approach the input signal frequency and, when this frequency modulation grows large enough, the PLL will lock on the input. Observation of v_c just before this happens will show how complex this behaviour really is. As the VCO free running frequency is approached, the waveform of v_c begins to depart radically from that of a sine wave, and develops a sharp cusp as the VCO frequency is pulled away from the input signal frequency. This is understandable because it is then that the rate of change of phase is at a maximum. The waveform of v_c is distorted in the opposite way, its curvature is softened, as the VCO moves towards the input signal frequency, and the rate of change of phase is reduced. As the input signal frequency gets closer and closer to the VCO free running frequency, the distortion of v_c gets worse, its amplitude grows, and a large statistical component, or jitter, develops. Then, suddenly, the display collapses to a simple horizontal line as the PLL drops into lock. The theoretical treatment of this complex phenomenon is very interesting, and the book by Best [1] is a good guide to the literature. One paper which should be mentioned is Verrazzani's generalised study of the problem [10], as his paper includes a picture of the complex capture transient, showing $d\varphi/dt$ as a function of time.

9.8 Improving the loop damping

The very long time constant, $C_2 R_3$, which has been used up to now for the experimental circuit, shown in Fig. 9.4, has made possible some interesting observations of loop behaviour, but the loop is far too lightly damped to be of much practical use. The only exception would be an application that called for a loop with a 'flywheel' action: a loop that could tolerate a very short break in the input signal and yet stay in lock.

Glancing back at the closed-loop transfer function for a PLL with a filter of this simple kind, equation (9.7), shows that reducing CR will increase the loop natural resonant frequency, equation (9.8), and also increase the damping. This follows because substituting equation (9.8) into equation (9.9) gives the result

$$\alpha_1^2 = 1/2\pi K_o K_d CR. \tag{9.10}$$

As K_o and K_d are more or less constant, once the VCO free running frequency has been chosen, it would seem at first sight that the best tactic would be to reduce CR until $\alpha_1^2 = 2$, this being the value that gives maximally flat response in a second order system. This would be quite wrong, however, because the value of $C_2 R_3$ that would make $\alpha_1^2 = 2$ is just below $0.6\ \mu s$. Such a small time constant would offer virtually zero attenuation to the double frequency component coming from the phase detector when the PLL is in lock. There would be a very large component of ripple on the VCO control voltage, v_c, and this would certainly interfere with its action. A compromise value of $C_2 R_3$, in between the very large value of 4.7 ms, that is being used at the moment, and the obviously too small value of $0.6\ \mu s$, that equation (9.10) suggests, will give improved loop damping, a larger capture range, and still give enough filtering or

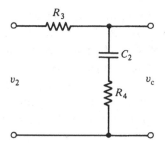

Fig. 9.6. *The lead–lag filter. The component labels correspond to those in the CD74HC4046A data sheet* [9], *and the input and output variables correspond to those shown in Fig. 9.3.*

smoothing on the phase detector output. Experiments along these lines are well worth while, but there is a better way to solve this problem, and that is to use a slightly more complicated filter. This is the kind shown in Fig. 9.6: the so-called lead–lag filter [11].

Some very simple algebra shows that the closed-loop transfer function for a PLL of the kind shown in Fig. 9.3, when the RC filter shown there is replaced with the filter shown in Fig. 9.6, is given by

$$\tilde{\varphi}_{out}/\tilde{\varphi}_{in} = \omega_n^2(1 + sC_2R_4)/(\omega_n^2 + \alpha_2\omega_n s + s^2) \qquad (9.11)$$

in contrast to equation (9.7), where

$$\omega_n^2 = 2\pi K_o K_d/C_2 R_3 \qquad (9.12)$$

as before, and

$$\alpha_2 = 1/\omega_n C_2 R_3 + \omega_n C_2 R_4 \qquad (9.13)$$

now gives a measure of the damping.

Using the earlier values of ω_n, C_2 and R_2, the value of R_4 which will make $\alpha_2 = \sqrt{2}$, the value that should make the loop have a maximally flat response, comes out to be $R_4 = 1.2\,k\Omega$. This should be checked experimentally by using an FM signal source. With sinusoidal, and very low level, audio frequency FM on an input signal carrier at 200 kHz, the demodulated output may be observed on pin 10 as the modulating frequency is increased towards $\omega_n/2\pi$, which is about 2 kHz. This demodulated output should be a constant, at constant low level modulation, as the modulating frequency is increased, rolling off at 12 db/octave above $\omega_n/2\pi$ with negligible peaking. When $R_4 = 0$, in contrast, there is a very large resonance at $\omega_n/2\pi$.

There is quite a large ripple on the VCO control voltage, v_c, with this lead–lag filter, and this will be seen as a rectangular wave, of varying mark to space ratio, and at double the VCO frequency, superimposed upon the signal on pin 10. With the component values suggested above, the ripple is about 100 mV peak to peak.

The ripple may have a very important effect upon the behaviour of the PLL in that it will be responsible for increasing the apparent capture range. As the addition of R_4 has had no effect upon the value of ω_n, adding R_4 would not be expected to change the capture range from the very small value observed when $R_4 = 0$. In fact, with $R_4 = 1.2$ kΩ, a capture range of nearly 100 kHz may be observed with some samples of the CD74HC4046A. This kind of behaviour is found in many PLLs and is by no means a simple phenomenon.

A clue as to what is happening may be obtained by repeating the capture observations described at the end of section 9.7. With $R_4 = 1.2$ kΩ, no beat frequency between the input signal and the free running VCO will be observed just before capture, only a very complicated and constant amplitude ripple voltage. Now open the loop, by disconnecting the low pass filter from pin 9, and take pin 9 to $+2.5$ V, provided by a simple potential divider of two 10 kΩ resistors in series from V_{cc} down to ground. The VCO will now run at its centre frequency of 200 kHz. If a ripple voltage is now added to this $+2.5$ V, by simply connecting a square wave generator to pin 9, through a capacitor of 0.1 μF, the explanation for the wide capture range, discussed above, will be clear. A 100 mV amplitude square wave, varied around a frequency of 400 kHz, is quite enough to pull the VCO frequency from its centre frequency by a considerable amount. The reason for this lies, of course, in the circuit detail of the VCO, and will be discussed in section 9.13.

In applications where the PLL is used as a demodulator, the ripple on pin 10 is removed by a further low pass filter. This filter is not in the loop and, consequently, has no influence upon the loop dynamics.

9.9 The type II phase detector

The second type of phase detector which is available in the CD74HC4046A PLL is shown in Fig. 9.7. This is an example of the so-called charge–pump type of phase detector, and, although this term does not have quite the same meaning here as it had in section 8.3 the name is well chosen [12].

What Fig. 9.7, in fact, shows is a positive edge triggered phase detector. Assume that the D-type flip-flops are initially in the RESET state, and let the signal input to pin 14 lead the VCO input to pin 3 by a small angle φ. The positive edge of the input signal will clock the upper D-type flip-flop, shown in Fig. 9.7, to have its $\bar{Q}$ output low. The lower D-type flip-flop is still reset with its $\bar{Q}$ output high. Following through the signals from these two $\bar{Q}$ outputs to the gate of Q_1 shows that this p-channel MOST will have its gate low and will thus be on. At this stage both inputs to the NOR gate,

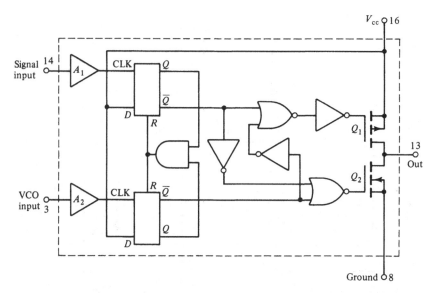

Fig. 9.7. Fig. 9.7. *A block diagram of the type II phase detector in the CD74HC4046A. The figure has been adapted from the data sheet. A_1 and A_2 are the same limiting amplifiers previously shown in Fig. 9.4.*

driving Q_2, are high, so that the gate of Q_2 is low, Q_2 is off, and the detector output, pin 13, is at V_{cc}.

When the positive going edge of the VCO clocks the lower D-type flip-flop, both flip-flops reset. Q_1 then goes off and Q_2 remains off. The output, pin 13, is now a very high impedance.

The opposite happens when the VCO leads the input signal. It is then Q_2 which turns on for the short time between the arrival of the positive edge on pin 3 and the positive edge on pin 14. During this time the output, pin 13, is held at ground. In fact, the output is a TRI-STATE (a trademark of the National Semiconductor Corp., as Horowitz and Hill point out [13]). Fig. 9.8 makes this clear: the type II phase detector in the CD74HC4046A may be represented by a three position switch which connects pin 13 to V_{cc}, or to ground, or to just an open circuit.

Fig. 9.8 is the second experimental circuit for this chapter, and shows the CD74HC4046A connected up to run with the type II phase detector, and with the kind of low pass filter that is needed. In fact, Fig. 9.8 is just a simple modification of Fig. 9.4: new values must be worked out for R_3, C_2 and R_4, and pin 13 used to input the low pass filter instead of pin 2. Everything else is as before.

To work out the correct values for R_3, C_2 and R_4, consider the loop to be in lock with an input signal at the VCO centre frequency, which is again

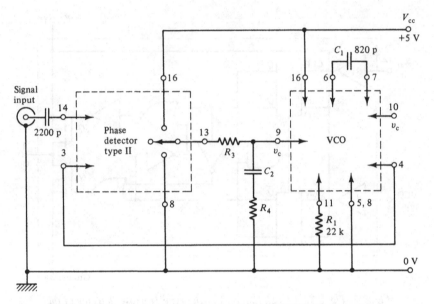

Fig. 9.8. *The experimental circuit using the CD74HC4046A with its type II phase detector.*

200 kHz. Under such conditions, $v_c = V_{cc}/2$, and the phase detector is an open circuit: the switch representing the phase detector TRI-STATE output is in the position shown in Fig. 9.8. $V_{cc}/2$ is stored on capacitor C_2, and remains there because of the very high input impedance to pin 9.

Now suppose that the input signal is increased very slightly in frequency. This means that the input signal now leads the VCO by the small angle φ, discussed above, and if the VCO does not follow, by also going to a higher frequency, φ will simply increase. The switch shown in Fig. 9.8, however, does not remain in the centre position but moves to the upper position for a time $(\varphi/2\pi)T$, where T is the period of the input signal.

When this happens, provided $R_4 \ll R_3$, the resistor R_3 has V_{cc} at its input end, pin 13, and $V_{cc}/2$ at its output end, pin 9. A current $V_{cc}/2R_3$ will thus flow for the short time, $(\varphi/2\pi)T$, that the switch remains in the upper position, and a charge, $(V_{cc}\varphi/4\pi R_3)T$, will be transferred to C_2, increasing the voltage across C_2 by a small amount. The frequency of the VCO will then increase, and the VCO will begin to move back into phase with the incoming signal.

The transfer function, $\tilde{v}_c/\tilde{\varphi}$, of the type II phase detector and its filter can be found by noting that the *mean* current flowing in R_3 must be $V_{cc}\varphi/4\pi R_3$. As v_c, the VCO control voltage, is simply the result of this current flowing through the impedance formed by C_2 and R_4 in series, it follows that

$$\tilde{v}_c/\tilde{\varphi} = V_{cc}(1 + sC_2 R_4)/4\pi sC_2 R_3. \qquad (9.14)$$

As φ may be taken as the input variable, φ_{in}, in the same way as in the previous analysis involving Fig. 9.2, and as the change in the VCO frequency, $\Delta f = d\varphi_{out}/dt = 2\pi K_o v_c$, it follows that

$$\tilde{\varphi}_{out}/\tilde{v}_c = 2\pi K_o/s \qquad (9.15)$$

which again expresses the integrating action of the VCO, noted above after equation (9.3).

Multiplying equations (9.14) and (9.15) then yields the open-loop transfer function for the PLL shown in Fig. 9.8. Using equation (9.6) then gives the closed-loop transfer function as

$$\tilde{\varphi}_{out}/\tilde{\varphi}_{in} = \omega_n^2(1 + sC_2 R_4)/(\omega_n^2 + \alpha_3 \omega_n s + s^2) \qquad (9.16)$$

which should be compared with equation (9.11): the transfer function for the PLL which used the type I phase detector with the same kind of filter now being used with the type II phase detector.

In equation (9.16), the loop natural resonant frequency is given by

$$\omega_n^2 = K_o V_{cc}/2C_2 R_3 \qquad (9.17)$$

while the loop damping is now given by

$$\alpha_3 = \omega_n C_2 R_4. \qquad (9.18)$$

In contrast to the PLL which uses the type I phase detector, this charge–pump PLL has no damping when $R_4 = 0$. This is clear when equation (9.18) is compared with the earlier result, equation (9.13).

9.10 Experiments with the type II phase detector

Comparison of equations (9.12) and (9.17), bearing in mind that K_d was equal to V_{cc}/π for the type I phase detector, shows that $C_2 R_3$ should be reduced to 25% of its previous value if the same value of ω_n is to be obtained with the type II phase detector. A suitable choice is $R_3 = 120$ kΩ and $C_2 = 0.01$ μF, making $C_2 R_3 = 1.2$ ms in contrast to the earlier 4.7 ms. Further calculations, using equation (9.18), show that $\alpha_3 = \sqrt{2}$, the condition for a maximally flat loop response, when $R_4 = 10$ kΩ.

Such a large value of R_4 may cause problems with this PLL. The loop may appear to lock over a very wide range, the same range of at least 100–300 kHz found previously for the type I phase detector, but close examination of the VCO and input signal waveforms may show that a

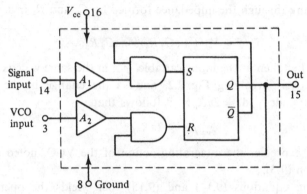

Fig. 9.9. *A block diagram of the type III phase detector in the CD74HC4046A. The figure has been adapted from the data sheet* [9]. *Again, A_1 and A_2 are the limiting amplifiers first shown in Fig. 9.4.*

considerable jitter exists between the two. The reason for this will be clear when the VCO control voltage is viewed, on pin 10, using a sensitive oscilloscope on a.c. Sharp voltage spikes of alternate polarity may be observed, showing that the VCO frequency is 'hunting' about the correct value. This is not because of incorrect system damping, but because the voltage spikes are interfering with the action of the VCO.

Reducing R_4 will cure this problem, at the expense of making the PLL rather lightly damped. A value of $R_4 = 2.2\,\text{k}\Omega$ is suitable. The same experimental program, described above for the type I phase detector, may then be followed. The capture range for this type II phase detector will be found equal to the lock range, and jitter, once in lock, should be really negligible. When the PLL is just on the point of capture, a lightly damped transient oscillation should be seen periodically on pin 10. This is the transient discussed by Gardener [12] and is shown in his paper as Fig. 10. Once in lock, the VCO control voltage is very free from ripple with this type II phase detector.

9.11 The type III phase detector

The third type of phase detector which is available on the CD74HC4046A is shown in Fig. 9.9. This is a positive edge triggered *RS* flip-flop, and it is clear that the output, from pin 15, will go to V_{cc} when the positive going edge of the signal input arrives, and then be reset to zero when the positive going edge of the VCO arrives. The initial state is assumed to be $\bar{Q} = 1$.

This type III phase detector thus needs a low pass filter in order to provide a VCO control voltage, v_c. This control voltage will be just above zero when the input leads the VCO by a small angle, increasing linearly towards V_{cc} and φ increases towards 2π. Should φ exceed 2π, v_c, will fall back at once to near zero: the output from this type III phase detector, when the PLL slips out of lock, is a sawtooth waveform, whereas the type I phase detector produces a triangle waveform.

It follows that the theory for a PLL using a type III phase detector is just the same as that for type I, except that K_d should be given the value $V_{cc}/2\pi$ instead of V_{cc}/π. When in lock, at the VCO centre frequency, $\varphi = \pi$ with the type III phase detector. Ripple on the VCO control voltage is more of a problem with the type III phase detector, because it is at the signal frequency, not double the signal frequency. The capture process with a type III phase detector is most interesting because, when the PLL is not in lock, the VCO free running frequency does not take up the value corresponding to $v_c = +2.5$ V, as it does with a type I phase detector. This should be looked at carefully. A further complication is that a PLL using a type III phase detector can lock on input signals having some harmonic relationship to the VCO. This is the same kind of behaviour, noted above, found with the type I phase detector.

9.12 Circuit shapes for phase detectors

The three types of phase detector in the CD74HC4046A are members of three classes into which virtually all analog phase detectors may be placed.

Classification of a phase detector as type I implies the exclusive OR function, which is identical to analog multiplication. When this is followed by a low pass filter, the detector has the characteristic given by equation (9.1) and the PLL will lock over the range $0 < \varphi < \pi$, always provided that both input and VCO signals have 50% duty cycles.

Classification of a phase detector as type II implies the charge–pump mode of operation. A capacitor is charged repetitively with very short pulses of current when the input signal is above the VCO frequency, and repetitively discharged when it is below. This type of phase detector does not need signals with a 50% duty cycle, and it ignores input signals which are harmonically related to the VCO. A real advantage of the type II phase detector is that its output is virtually ripple free when the loop is in lock.

Classification of a phase detector as type III implies the positive edge triggered set–reset mode of operation. After low pass filtering, this phase detector provides an output which varies linearly over the range $0 < \varphi < 2\pi$. It can handle inputs of any duty cycle.

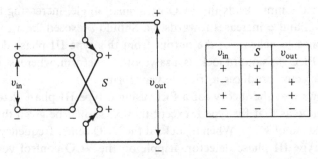

v_{in}	S	v_{out}
+	+	−
−	+	+
+	−	+
−	−	−

Fig. 9.10. *A simple change-over switch effects the exclusive OR function. Switch positions are labelled + and −, while the same labels are used to define the polarity of input and output signals.*

These three classes of phase detector may now be considered from the point of view of their circuit shapes. The exclusive OR function, which forms the basis of the type I phase detector, may be realised with the simple change-over switch circuit shown in Fig. 9.10 When the switch positions are labelled + and −, and the same symbols are used to define the polarity of the input and output, the truth table shown in Fig. 9.10 confirms the exclusive OR property of this simple circuit.

The same change-over switch topology, or circuit shape, is found in a well-known four quadrant multiplier circuit, first described in 1968 by the simultaneous publications of Bilotti [14] and Gilbert [15]. This circuit is shown in Fig. 9.11, and, again, the adjoining truth table shows the 'if only v_1 or only v_3 is positive, then v_2 is positive' function essential to the exclusive OR. The reason for this circuit property, however, lies in the cross connection of Q_3, Q_4, Q_5 and Q_6. These four transistors play the same role as the switch contacts in Fig. 9.10.

The circuit shown in Fig. 9.11 is at the heart of many integrated circuit analog multipliers [16], and its use as a phase detector in an early PLL has been dealt with in detail by Gray and Meyer [17].

A second example of a type I phase detector, which has the same circuit shape as the simple change-over switch shown in Fig. 9.10, is shown in Fig. 9.12. Four diodes are used in this case as the four contacts which are needed to implement a change-over function. This is quite an old idea in electronics, often referred to as a ring demodulator [18].

The particular circuit shown in Fig. 9.12 would have a VCO signal of several volts applied to the v_3 input, but the signal input to v_1 would be small compared to a forward diode drop. The circuit then functions with D_1 and D_3 turning on during the positive half cycle of the VCO, while D_2 and D_4 turn on during the negative half cycle. When there is no

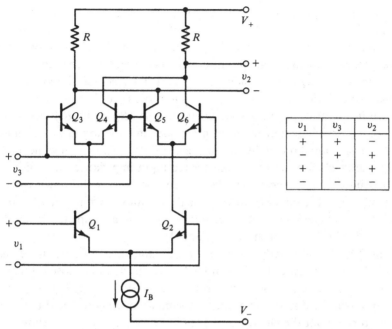

v_1	v_3	v_2
+	+	−
−	+	+
+	−	+
−	−	−

Fig. 9.11. *The four quadrant analog multiplier circuit which is often employed as a type I phase detector.*

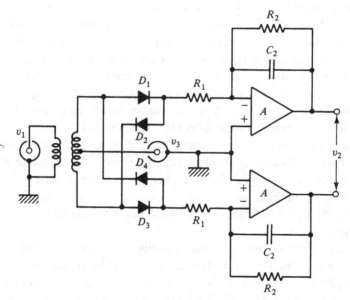

Fig. 9.12. *A schematic circuit diagram of a type I phase detector using diodes in a change-over switch mode. As in Fig. 9.1, v_1 would be the signal input, v_3 the VCO, and v_2 the output.*

component of v_1 in phase with v_3, the output, v_2, will be zero, because the demodulator is balanced. When v_1 is in phase with v_3, the output will be negative, and when v_1 lags or leads v_3 by π, the output will be positive. The actual value of the differential output depends upon the turns ratio of the input transformer, which would usually be $1:1+1$ because it would be a trifilar wound wide-band transformer on a toroid core. The two amplifiers shown in Fig. 9.12 have transfer functions $-R_2/R_1(1+sC_2R_2)$, the same low pass filter characteristic given previously in equation (9.4), but now with the additional possibility of some gain: R_2/R_1. The actual value of R_1 may also be chosen to provide the desired input impedance for v_1 and v_3.

It is interesting to note that there is another very well-known circuit [19], of exactly the same *circuit shape* as the circuit shown in Fig. 9.12, and this is also a type I phase detector but it works on a completely different principle. This circuit is shown in Fig. 9.13.

Fig. 9.13 shows exactly the same 'ring' connection of four diodes as Fig. 9.12: in both cases the four diodes are connected in a closed ring, all forward conducting in the same direction around this ring. Fig. 9.12 was drawn in such a way that the ring was somewhat obscured, and this was done to bring out the change-over switch topology. The circuit shown in Fig. 9.13 does not operate like a change-over switch at all. The high level input to this phase detector, the local oscillator, v_3, is now applied so that *adjacent* diodes, D_1 and D_2 followed by D_3 and D_4, are turned on every half cycle. In the circuit shown in Fig. 9.12, diodes *opposite* one another are turned on: D_1 and D_3 in one half cycle, D_2 and D_4 in the text.

The local oscillator level, in the circuit shown in Fig. 9.13, is set so that the forward current in the diodes makes them present an impedance at the local oscillator input socket which is a reasonable match. The signal input transformer then has a secondary load consisting of two forward biassed diodes and two reverse biassed diodes in a series–parallel combination. As the forward biassed diodes will be mainly resistive, and the reverse biassed diodes mainly capacitive, this load impedance may be designed to give a good match at the signal input socket, over quite a wide range of frequency, by making use of the transformer leakage inductance to resonate with the diode capacitance. It follows that this circuit is used for VHF applications.

The output from the circuit shown in Fig. 9.13 comes from the centre tap of the local oscillator transformer. There will be no output when the input signal is at the local oscillator frequency, and also lagging or leading the local oscillator by $\pi/2$. This follows because the circuit will be balanced under these conditions. A signal input at a frequency other than the local oscillator frequency, or one which has a component in phase with

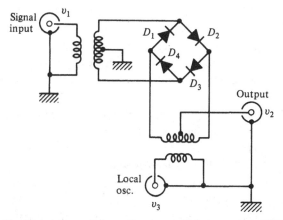

Fig. 9.13. *The ring, or balanced, mixer which is used at VHF. This works on quite a different principle to the previous circuit, shown in Fig. 9.12, although it has the same circuit shape.*

the local oscillator, will produce an output because it will unbalance the circuit. For example, if D_1 and D_2 were forward biassed by the local oscillator, a signal input which was in phase with the local oscillator would decrease the current in D_1 but increase the current in D_2. Then D_1 would have a higher resistance than D_2, because of the non-linear forward characteristic of these diodes, and an output would result. This output would be at the double frequency, plus a constant level, characteristic of a type I phase detector. The two circuits, shown in Figs. 9.12 and 9.13, are both type I phase detectors, and both have the same circuit shape, but they work on quite different principles, because the signal levels and the input and output connections are quite different. This is something to look out for in electronic circuit design.

Turning now to the type II, or charge–pump type, phase detector, the schematic block circuit diagram, Fig. 9.7, shows how this kind of phase detector may be realised with standard logic gates and flip-flops. The TRI-STATE output circuit, shown in Fig. 9.7, is also a standard feature in all logic families.

An interesting paper by Yaeger [20] describes a charge–pump phase detector using bipolar transistors, in contrast to the MOSTs of Fig. 9.7, but the logic driving these bipolar devices is virtually the same in both designs. Yaeger's paper is particularly useful in that it tackles the problem of compensating for the changing closed-loop transfer function of a PLL when this is used in a frequency synthesizer. Frequency synthesis with a PLL involves a high frequency VCO followed by a frequency divider [2]. The phase detector deals with the lower frequency output of the divider, which is made to lock on to a low frequency crystal oscillator. As the division ratio is changed, to obtain different output frequencies, so does the effective value of K_o, and hence the loop gain of the PLL. Type II phase detectors are a good choice for frequency synthesizer applications, because of the wide capture range which is possible. Yaeger adds a further advantage, by slight modification of the TRI-STATE output circuit, which enables the current pulse it supplies to be varied in magnitude, and thus compensate for the changes in K_o, mentioned above.

Finally, type III phase detectors may be considered, along with a few other circuit ideas for phase detectors which have been proposed.

Fig. 9.9 shows a schematic block diagram of the type III phase detector used in the CD74HC4046A. This is implemented with standard logic gates and flip-flops. The whole point of the type III phase detector, as was summarised at the beginning of this section, is its ability to handle a variation in phase over the full range of zero to 2π. This means that the output from the low pass filter, which must be used with this type of phase detector, is a saw-tooth wave when the phase angle is increasing continuously, at a sufficiently slow rate for the low pass filter to follow, whereas the output from a type I phase detector, under the same conditions, is, of course, a triangle wave. Response over the full range, 0–2π, has advantages in some applications, and an interesting way of realising such a type III phase detector, which is quite different to that shown in Fig. 9.9, has been published by Fyath [21].

There are a number of other circuit ideas for phase detectors which have been published. These may be simplifications of some of the ideas which have already been dealt with here. For example, a phase detector identical to the one shown in Fig. 9.12, but with diodes D_2 and D_4 omitted, is a classical circuit shape for a phase detector [22]. Such a 'half wave' realisation has obvious disadvantages. Another apparent simplification of one of the circuits which has been treated above is the phase detector published by Soyuer and Meyer [23]. This is identical to the circuit shown in Fig. 9.11, except that Q_5 and Q_6 are omitted, and the collector of Q_2 is taken directly to V_+. The output from the circuit, v_2 in Fig. 9.11, will then be zero when v_1 is negative: the v_2 column in the truth table shown in Fig. 9.11 becomes $-0+0$, instead of $-+-$, as it is with the completely balanced circuit. Soyuer and Meyer were able to show that such a phase detector would have better dynamic performance, in a PLL with a square wave VCO, for both deterministic and random input signals.

The review of phase detectors, which has been given in this section, is incomplete and has been restricted to circuits which are really analog phase detectors. No mention of the fully digital phase detectors, considered briefly in section 9.2, has been made in this section, but these are of great importance in digital PLLs, and the reader is again referred to Best's work on this topic [7].

9.13 Circuit shapes for voltage controlled oscillators

The VCO in the CD74HC4046A, the device that was used for the experimental work in this chapter, is built up from the standard logic blocks which are available in CMOS. Fig. 9.14 shows the circuit detail of the VCO from this logic block point of view. The pin numbers in Fig. 9.14, and the amplifier A_4, are the same as in Fig. 9.4, where the complete CD74HC4046A was shown. This circuit detail may be inferred from the data sheet [9] and from an excellent application note [24] which was written for an earlier CMOS PLL: the CD4046A.

The action of the VCO shown in Fig. 9.14 may be understood, and examined experimentally, if pin 9 is taken to $V_{cc}/2 = +2.5$ V, pin 5 is grounded, to put the clock input of the *JK* flip-flop up at $V_{cc} = +5$ V and enable the VCO, and $C_1 = 820$ pF and $R_1 = 22$ kΩ as before. The amplifier A_4 then establishes a current $V_{cc}/2R_1$ (about 115 μA) in Q_1.

The oscillator part of the circuit shown in Fig. 9.14 is made up of Q_3–Q_6, C_1 and the *JK* flip-flop. The flip-flop ensures that either Q_3 and Q_6 are on (and Q_5 and Q_4 off), or that Q_5 and Q_4 are on (and Q_3 and Q_6 off). In the former case, pin 7 will be held just above ground while pin 6 will rise at a constant rate, $dv/dt = I/C_1$, where I is the current in Q_2. Now Q_1 and Q_2 form a current mirror, but the area of Q_2 is about five times that of Q_1. This means that the current, I, available to charge up C_1, is about 500 μA, so that I/C_1 will be about 0.6 V/μs when $C_1 = 820$ pF.

The voltage on pin 6 will be observed to rise linearly at about this rate until it reaches a level close to 1.4 V. The K input on the *JK* flop-flop is then triggered, resetting $\bar{Q}$ high and Q low. This turns on Q_4 and turns off Q_6. The current in Q_2 is now directed, though Q_5, into C_1 again, but in the opposite direction to that of the previous situation. The waveforms actually observed are sketched in Fig. 9.15.

Fig. 9.15 shows that, during the time that Q_6 is on, pin 7 is not exactly at zero but positive by a small amount, due to the current in Q_6 causing a small drop across it. The same applies when it is the turn of Q_4 to be on: pin 6 is slightly positive. What may be unexpected in Fig. 9.15 is the asymmetry of the waveforms: if C_1 charges up to the $+1.4$ V shown in Fig. 9.15, why does the waveform not begin at -1.4 V and pass through zero at the quarter period? The answer to this question lies in the structure of Q_4 and Q_6 in the CMOS process. The drains of these two n-channel

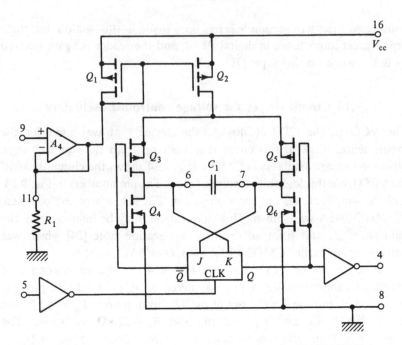

Fig. 9.14. *The essential circuit detail of the VCO in the CD74HC4046A.*

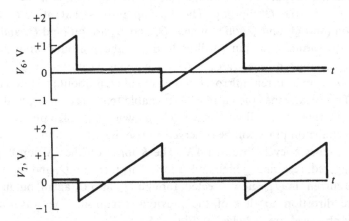

Fig. 9.15. *The waveforms that will be observed on pins 6 and 7 of the CD74HC4046A.*

MOSTs must be n-type regions of silicon in a p-type substrate which, as Fig. 9.14 shows, is grounded. It follows that this np intrinsic diode will 'catch' if any attempt is made to drive the drain below about -0.7 V. The intrinsic diode discharges C_1 a certain amount each half cycle, and this is another factor (the first being the current multiplication of Q_1 and Q_2) that makes this VCO operate at a higher frequency than would be expected from a first glance at the circuit and a note of the values chosen for C_1 and R_1. For this reason it is interesting to experiment with smaller values of C_1 and R_1, taking the operating frequency towards the 18 MHz maximum given on the data sheet, and observing and understanding the changes which occur in the waveforms.

At the end of section 9.8, attention was drawn to the problem of the ripple on the VCO control voltage that may have to be tolerated in some PLL designs. In an experiment, which was suggested in section 9.8, an externally generated square wave ripple voltage was injected at pin 9, on a free running VCO in a CD74HC4046A, and this was found to pull the oscillator frequency.

The circuit detail shown in Fig. 9.14 suggests why a ripple voltage on pin 9 might influence the VCO in this way. A sudden increase in the voltage at pin 9 would produce only a discontinuity in the *slope* of the ramp across C_1, if Q_4 and Q_6 had really zero resistance, and if C_1 were really a perfect capacitor. This is not the case, however, and a step increase in the voltage at pin 9 must produce a step increase, or, even more important, a voltage spike, at pins 6 or 7. This may trigger the *JK* flip-flop early, and thus increase the operating frequency so that the VCO locks on to the ripple voltage. Similarly, a step decrease in the voltage at pin 9 may produce synchronisation at frequencies below the free running frequency. All these possibilities may be examined experimentally with the CD74HC4046A. It is important to note that the same problem arises with virtually all VCO designs.

Fig. 9.16 shows a VCO circuit which is a classical circuit inasmuch as it was used in a whole series of very successful bipolar PLLs, the NE560, 561 and 562, introduced by Signetics in 1971 [25]. This circuit, which is dealt with in detail by Gray and Meyer [26], has much in common with the previous VCO, shown in Fig. 9.14. In both circuits, the capacitor, C_1, is being charged with a constant current, first from one side and then the other. The output from the VCO shown in Fig. 9.16 is the square wave developed across R_1 and R_2, the voltage level there being clamped by Q_1 and Q_2.

This rather crude voltage clamping is replaced by much better technique in the current bipolar PLL of the 560 series: the NE564 [27]. This device, which can operate up to 50 MHz, uses simple resistors in place of $Q_1 R_1$ and $Q_2 R_2$, shown in Fig. 9.16, and there is an arrangement of current sinks which maintains a constant current in these resistors as the capacitor current is varied to change frequency. This results in a great improvement in the temperature stability of the VCO centre frequency. An important paper by Gilbert [28] reviews developments in this field, and lists the

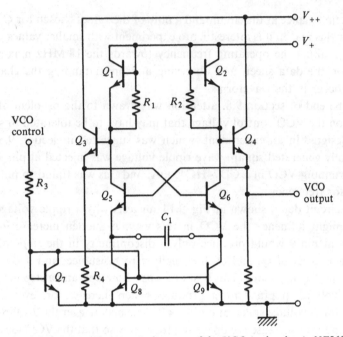

Fig. 9.16. *A simplified circuit diagram of the VCO in the classic NE560 PLL.*

advantages of the astable multivibrator circuit, of which Figs. 9.14 and 9.16 are examples. Gilbert's solution, to the temperature dependence problems of this circuit, is to use a subtle arrangement of a bandgap reference and diode switches. His paper is a good guide to the foundation literature in this area of electronic circuit design.

Developments in integrated circuit PLLs have taken the technique into the VHF region. A device fabricated with a 2 μm CMOS process can be used up to 128 MHz [29] and the device described by Soyuer and Meyer [23], referred to in the previous section, can operate up to 350 MHz. This latter device was fabricated with a 2 μm bipolar process, and was interesting in that it used a varactor tuned VCO with a high Q LC resonator. This meant an external inductor had to be used with this device, but, because of the high Q possible with this technique, a very small noise bandwidth for the resulting PLL could be obtained.

9.14 Conclusions

This chapter began with a review of PLL theory, which was supported by a series of experimental circuits built around the CD74HC4046A device. This made it possible to introduce the three most commonly used kinds of phase detector found in analog PLLs and discuss the low pass filters which must be used with these different kinds of phase detector.

In all these cases, the behaviour of the PLL is best examined by looking at the way in which the VCO control voltage changes as the frequency of the input signal is varied. Fig. 9.5 shows one particular example. Measurements along these lines show up all the complicated behaviour of a PLL as it drops into lock, either with the expected signal, or with one which may be quite unexpected in that it has some harmonic relationship to the VCO frequency at that particular instant.

The charge–pump, or type II, phase detector is perhaps the most interesting one to be looked at here. It provides a very wide capture range, and, when the low pass filter is designed correctly, the PLL is very stable and jitter free. A further advantage of the type II phase detector is that it makes the loop sensitive only to input signals which lie within the tuning range of the VCO.

The chapter continued with a look at the circuit detail, the circuit shapes and circuit ideas, found in phase detectors and VCOs. In the case of phase detectors, circuit shapes have remained very much the same over a very long period of time. Developments in this area have clearly been in device performance: higher speeds and better device matching, both the result of advances in process technique.

Circuit development for VCOs appears to be more circuit shape, or circuit idea, directed. This is clear when the early 560 circuits [25] are compared with the far more complex 564 circuit [27], or with a more recent device [29]. All use the basic astable multivibrator circuit shape, shown in Figs. 9.14 and 9.16, but advances have been made by adding compensating circuits, or more accurate level references. There have also been new developments in the way that the VCO problem may be tackled, which have not been discussed here in connection with PLLs because the VCO for a PLL must have a very fast response to a change in its input control voltage. These new developments have been called 'precision charge dispensing' techniques by Gilbert [28], and are used, for example, in the AD651 voltage to frequency converter [30]. The charge dispensing technique is similar to the method described here, in chapter 8, to obtain an ultra-linear VCO. This was done by using a charge–pump frequency to voltage converter as a feedback element across a crude VCO, and resulted in the experimental circuit shown in Fig. 8.6. Such an ultra-linear VCO, which also has remarkable stability against changes in temperature, does not have the speed of response called for in most PLL applications.

Notes

1 Best, R. F., *Phase-Locked Loops: Theory, Design and Applications*, McGraw Hill, New York, 1984.

2 Egan, W. F., *Frequency Synthesis by Phase Lock*, John Wiley, New York, 1981.

3 de Bellescize, H., *L'Onde electrique*, **11**, 209–72, 1932.

4 Tucker, D. G., and Ridgway, J. F., *Electronic Engineering*, **19**, 75–6, 238, 241–5, 276–7, 366 and 368, 1947.

5 Horowitz, P., and Hill, W., *The Art of Electronics*, Cambridge University Press, Cambridge, second edition, 1989, pp. 641–55.

6 Gray, P. R., and Meyer, R. G., *Analysis and Design of Analog Integrated Circuits*, John Wiley, New York, second edition, 1984, pp. 605–34.

7 Note 1 above, pp. 89–96.

8 Note 6 above, p. 528.

9 RCA Data File No. 1854.

10 Verrazzani, L., *IEEE Trans. Aero. Electr. Syst.*, **AES-14**, 329–33, 1978.

11 Note 5 above, p. 648.

12 The term may have come into wide use after the publication of F. M. Gardener's review: Charge–pump phase-lock loops, in *IEEE Trans. Comm.* **COM-28**, 1849–58, 1980. Gardener is also author of the classic *Phaselock Techniques*, John Wiley, New York, second edition 1979.

13 Note 5 above, p. 487.

14 Bilotti, A., *IEEE J. Sol. St. Circ.*, **SC-3**, 373–80, 1968.

15 Gilbert, B., *IEEE J. Sol. St. Circ.*, **SC-3**, 365–73, 1968.

16 For example, the Analog Devices AD530 series, the Motorola MC1459L, and the RCA CA3091D.

17 Note 6 above, pp. 622–32.

18 For example, by Tucker and Ridgeway, who used the ring demodulator as a phase detector in one of their circuits back in 1947. See note 4 above, p. 277.

19 Skolnik, M. J., *Radar Handbook*, McGraw-Hill, New York, 1970, p. 5.42.

20 Yaeger, R., *RCA Review*, **47**, 78–87, 1986.

21 Fyath, R. S., *Electronic Engineering*, **59**, No. 726, 28, 1987.

22 Greenwood, I. A., Holdam, J. A., and Macrae, D., *Electronic Instruments*, McGraw-Hill, 1948, p. 384.

23 Soyuer, M., and Meyer, R. G., *IEEE J. Sol. St. Circ.*, **SC-24**, 787–95, 1989.

24 The RCA COS/MOS phase-locked-loop. A versatile building block for micro-power digital and analog applications. ICAN 6101. In: *COS/MOS Integrated Circuits*, RCA Solid State Data Book, SSD 250C, 1983, pp. 714–17.

25 Grebene, A. B., *IEEE Trans. Broadcast and Television Receivers*, **BTR-17**, 71–80, 1971.

26 Note 6 above, pp. 628–31.

27 For the circuit diagram of this device, see the Data Brief in *Radio and Electronics World*, pp. 49–51, Nov. 1983.

28 Gilbert, B., *IEEE J. Sol. St. Circ.*, **SC-11**, 852–64, 1976.

29 Kato, K., Sase, T., Sato, H., Ikushima, I., and Kojima, S., *IEEE J. Sol. St. Circ.*, **SC-23**, 474–84, 1988.

30 *1986 Update and Selection Guide*, Analog Devices Inc., Massachusetts, pp. 3.137–3.152.

10
Low noise circuits

10.1 Introduction

As Wilmshurst [1] has written, noise in electronics has, today, come to mean 'almost any kind of unwanted signal in an electronic system'. This is in contrast to the classical picture of noise as being a problem area which is only concerned with the fact that electronic circuits operate at a finite temperature and also have to operate with electric currents that are really made up of a flow of discrete charged particles. Further evidence for the wider view which is now taken of noise problems in electronics can be taken from the use of the term 'electromagnetic compatibility (EMC) [2]'.

For the above reason, this chapter is really in two parts. To begin with, circuit shapes and circuit ideas that attempt to minimise the effects of the intrinsic thermal and shot noise of electronic devices will be considered. This calls for a brief summary of some well-known theory which will be given first. The topic then divides fairly naturally into low and high frequency amplifiers, and some interesting experimental circuits can be proposed for both fields. After this look at these classical kinds of noise problems, the chapter concludes by considering some of the circuit ideas which have been proposed to eliminate very special noise problems in various signal processing systems.

10.2 Intrinsic thermal noise sources

An excellent reference for the fundamentals of noise in electronic circuit design is chapter 11 of the book by Gray and Meyer [3]. For a deeper treatment of noise in solid state devices, both Bell [4] and Buckingham [5] will be found valuable.

The origin of thermal noise, not only in electronic systems, is the finite temperature and the discrete particle nature of the world we live in. There must be a noise power, kTB, associated with any signal channel, which is added to the signal power that channel may carry. Here k is the Boltzmann constant, 1.38×10^{-23} J/K, T is the absolute temperature in Kelvin, and B is the channel bandwidth in hertz. For audio frequency work, kTB is extremely small, and is, of course, a measure of the sensitivity of the human ear. Taking T to be 290 K and B to be 10 kHz, $kTB = 4 \times 10^{-17}$ W. In this example, the noise is due to the finite temperature of the air and the fact that air is made up of discrete molecules. The remarkable sensitivity of the human ear, estimated above, can be confirmed experimentally [6].

Applied to electrical systems, this fundamental idea of a noise power source, kTB, associated with any signal channel, leads to the result that any resistor, value R, may be represented by a noiseless resistor in series with a thermal noise voltage source of rms value

$$e_n = (4kTBR)^{\frac{1}{2}}. \qquad (10.1)$$

Alternatively, the representation may be a noiseless resistor in parallel with a thermal noise current source of rms value

$$i_n = (4kTB/R)^{\frac{1}{2}}. \qquad (10.2)$$

The rms noise voltage, given by equation (10.1), and the rms noise current, given by equation (10.2), both increase with the square root of the system bandwidth. This follows, of course, from the fact that the available noise power, kTB, increases with bandwidth directly. This is why the units nV/Hz$^{\frac{1}{2}}$ and pA/Hz$^{\frac{1}{2}}$ are frequently used in noise data as measures for e_n and i_n. An actual measurement of noise is nearly always a power measurement (the spectral density) taken over a well-defined bandwidth, and then this power, referred to unit bandwidth for convenience, is converted into an equivalent voltage or current.

10.3 Intrinsic shot noise sources

The fact that an electrical current is really a flow of discrete charges means that there must be a statistical component associated with any current. This means that the collector current, or drain current, of a transistor must have a shot noise component. For the bipolar transistor there will also be a shot noise component associated with the base current, and a thermal noise source associated with the unavoidable bulk resistance of the base.

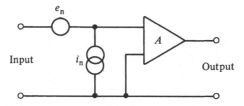

Fig. 10.1. *The equivalent circuit for noise in a solid state amplifying device. The ideal device, A, is noise-free and the noise of the real device is all referred to the input where it is represented by the two generators* e_n *and* i_n.

At low frequencies, the audio frequencies, the noise in solid state devices can become much greater than the simple shot noise model would predict. This will be discussed briefly in the next section. For the moment, it is only necessary to note that it has become standard practice to represent the noise in a solid state device by means of the equivalent circuit shown in Fig. 10.1. Using this simple model, it is easy to show [7] that there is an optimum source impedance

$$R_{s(opt)} = e_n/i_n \tag{10.3}$$

which should be used with the device to give a minimum noise figure

$$F_{min} = 10 \log_{10} (1 + e_n i_n/2kT). \tag{10.4}$$

Noise figure is a measure of how much noise an amplifying device adds to the thermal noise which is intrinsic to the signal channel anyway, and this would be zero for a noiseless amplifier. Equation (10.4) shows that the really important noise parameter for an amplifying device is the *product* of the two generators, e_n and i_n, shown in Fig. 10.1. This product gives a measure of the amplifier noise power per unit bandwidth, while kT gives a measure of the thermal noise power over the same unit bandwidth.

10.4 Low frequency noise and the integrated circuit process

Very remarkable improvements in the noise performance of solid state devices have been achieved since the mid-1960s. These improvements are clearly due to far better processing technique.

Considering bipolar devices first, the main problem with the early devices of the 1960s was so-called $1/f$ noise, or flicker noise, which plagued circuit performance in the audio frequency range. Fig. 10.2 illustrates the problem. If the noise spectral density of a device is measured at its output, and then referred back to the input, as shown in Fig. 10.1, to relate to the

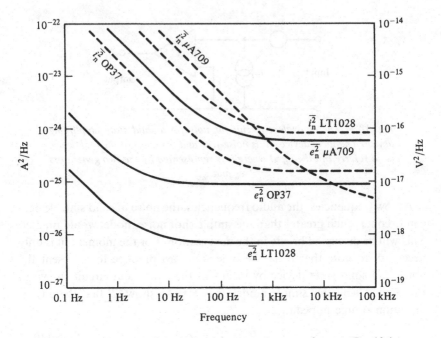

Fig. 10.2. *The spectral densities of the two generators shown in Fig.* 10.1
for a classical op-amp of 1965, *the* μA709, *for a low noise op-amp of*
1981, *the* OP37, *and for a more recent ultra-low noise op-amp, the*
LT1028.

noise sources $\overline{e_n^2}$ and $\overline{i_n^2}$, a $1/f$ variation is always found when measurements
are taken at a low enough frequency. What has changed dramatically
since the mid-1960s, is the frequency at which this $1/f$ noise, or flicker
noise, sets in. Improvements in processing technique have pushed this
frequency lower and lower, as well as reducing the noise overall.

The μA709 was an early high gain bipolar operational amplifier, first
described in 1965 [8]. As Fig. 10.2 shows the spectral density $\overline{i_n^2}$ never really
flattens out with this device as measurements are taken over the audio
range. Even at 100 kHz, $\overline{i_n^2}$ is still falling quite rapidly. In contrast, $\overline{e_n^2}$ for
the μA709 may be treated as a constant from 1 kHz to 100 kHz, but it
shows the $1/f$ behaviour at frequencies below 1 kHz.

It is this poor low frequency noise performance which more recent
developments in integrated circuit processing, and design, have put right.
Fig. 10.2 shows $\overline{e_n^2}$ and $\overline{i_n^2}$ for a well-established low noise operational
amplifier: the OP37 [9]. For the OP37, both $\overline{e_n^2}$ and $\overline{i_n^2}$ are really constant
down to 1 kHz, and $\overline{e_n^2}$ does not begin to show the $1/f$ variation until the
measurement frequency is below 10 Hz. A more recent ultra-low noise
bipolar operational amplifier, the LT1028 [10], is very similar as far as the

Table 10.1. *Three recent op-amps, the OP37, LT1028 and OP15, are compared with the classical μA709 for low frequency noise performance.*

Device	Type	Test freq. (kHz)	e_n (nV/Hz$^{\frac{1}{2}}$)	i_n (pA/Hz$^{\frac{1}{2}}$)	$R_{s(opt)}$	Noise (fig. dB)
μA709	Bipolar	1	7.75	0.71	10.9 kΩ	2.27
OP37	Bipolar	1	3.0	0.4	7.5 kΩ	0.61
LT1028	Bipolar	1	0.85	1.0	850 Ω	0.44
OP15	JFET	1	15.0	0.01	1.5 MΩ	0.08

frequency dependence of $\overline{e_n^2}$ and $\overline{i_n^2}$ is concerned. What is special about the LT1028 is its remarkably low $\overline{e_n^2}$.

Table 10.1 attempts to summarise these results. Going back to equations (10.3) and (10.4), the values of optimum source impedance and the resulting noise figure may be calculated for the three bipolar devices which were the subject of Fig. 10.2. These calculations are made at a frequency of 1 kHz, where the modern devices are showing constant levels of $\overline{e_n^2}$ and $\overline{i_n^2}$, an essential requirement for equations (10.3) and (10.4) to be valid. The calculations for the μA709 are only included for a rough comparison. The results shown in Table 10.1 indicate the very low noise figures which should be obtained with the OP37 and the LT1028 when these devices are used as audio amplifiers. The values obtained for $R_{s(opt)}$ are also interesting because a low value of $R_{s(opt)}$ can be a very useful feature. This follows because most signal sources, like microphones, sensors, magnetic recording heads, etc., have quite low impedance, and, even when this is not so, the signal source may have to be connected to its amplifier through a fairly low impedance cable.

Looking again at equations (10.3) and (10.4) shows that a good design strategy for a low noise operational amplifier is to aim to minimise the product $i_n e_n$, and thus minimise the noise figure given by equation (10.4), and also try to reduce e_n as much as possible, without being concerned if this implies an increase in i_n, because this will then reduce e_n/i_n and so reduce the value of $R_{s(opt)}$. Precisely this strategy has been adopted by the designers of the LT1028: the collector current in the input devices is made quite high, giving good wide-band performance, low e_n, but quite high i_n.

Not all amplifiers work from a low signal source impedance, however, and that is why Table 10.1 includes the OP15, an operational amplifier which has JFET input devices. The JFET is the best solid state device of all for very low frequency work because the $1/f$, or flicker, noise is very

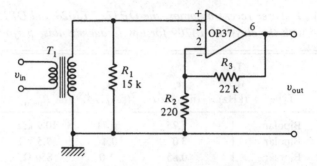

Fig. 10.3. *An experimental low noise audio pre-amplifier. The OP37*
operates on $\pm 15\,V$ supplies to pins 7 and 4, which must be well
decoupled to ground. T_1 is a high performance audio transformer (STC-
66121B or RS-217-804) connected 1:12.9 by wiring the two primaries
in parallel and the two secondaries in series. The transformer should be
mounted in its mu-metal can.

low [11]. As Table 10.1 shows, the JFET has a very low noise figure indeed
for audio work, but a very high value of $R_{s(opt)}$. An amplifier like the OP15
is thus ideal as a pre-amplifier for a crystal microphone, where it would be
mounted inside the microphone itself, or for the input stage of an
oscilloscope amplifier of the kind discussed in section 6.4. Working from
a low source impedance, however, the OP15 would appear to have high
noise, compared with a bipolar device, because of its very high value of e_n.

10.5 An experimental audio pre-amplifier

If a low noise operational amplifier like the OP37 is to be used as an audio
pre-amplifier, its optimum source impedance for minimum noise figure,
given in Table 10.1 at 7.5 kΩ, is unlikely to be suitable directly. For
example, a dynamic microphone will present a source impedance of
perhaps 100 Ω while a crystal microphone will present a source impedance
far higher than 7.5 kΩ.

Fig. 10.3 shows the kind of circuit which will overcome this problem
and cause the OP37 to work from its optimum source impedance of
7.5 kΩ. This is done by using a step-up input transformer with a secondary
load of $2R_{s(opt)}$. The input to the amplifier then presents an input
impedance of $2R_{s(opt)}/N^2$, where N is the turns ratio of the transformer.

Fig. 10.3 is the first experimental circuit for this chapter. T_1 must be a
high quality audio transformer, because magnetic components can have
noise problems themselves at very small signal levels, due to the properties
of the magnetic material used for the core [12], and also because of their

sensitivity to stray fluctuating magnetic fields. The transformer in the circuit shown in Fig. 10.3 provides some voltage gain, $N = 12.9$ in this case, and a further voltage gain of 100 is provided by the OP37 with its feedback network consisting of R_2 and R_3. Gray and Meyer deal with the problem of the additional noise that comes from the feedback resistors in a case like this [13], and show that the parallel combination of R_2 and R_3, in Fig. 10.3, is effectively added to the signal source impedance. For this reason, R_2 and R_3 are kept low. Another reason for keeping R_3 low is to exploit the very wide-band properties of the OP37. Connected as a feedback amplifier with a gain of 100, the OP37 should give this gain up to well over 100 kHz, provided stray capacitance across R_3 does not begin to increase the feedback.

The source impedance presented to the OP37 is 7.5 kΩ because, looking back into R_1, the 15 kΩ shown in Fig. 10.3, this is seen to be in parallel with $N^2 R_s$, where R_s is now the source impedance connected to the true input terminals, v_{in}. When $N = 12.9$, this true source impedance should be 90 Ω: a typical dynamic microphone impedance.

Now equation (10.1) shows that the rms noise voltage which should be measured across a 90 Ω resistor at 290 K, with a bandwidth of 100 kHz, is just under 0.4 μV. As the circuit shown in Fig. 10.3 has a total voltage gain of 1290, it follows that a noise voltage of about 0.5 mV rms should be observed at the output, if the circuit itself adds no additional noise. In fact, the experimentalist may well see a lower noise level at the output, when the input is simply connected to a 90 Ω resistor, because the overall bandwidth of the circuit shown in Fig. 10.3 will almost certainly be less than 100 kHz because it will be limited by the transformer to be about 50 kHz.

The noise output of 0.5 mV is at a level where it may be conveniently observed on an oscilloscope, when this is on its most sensitive range: 2–5 mV/div. The noise will be seen as a diffuse band across the screen with no random spikes or granularity. This is characteristic of a low noise device like the OP37. If the OP37 is replaced with an ordinary op-amp, like a 741, with which it is pin compatible, a difference in the appearance of the noise will be seen. Measured with an rms voltmeter, the noise at the output of the circuit with a 741 may well be less than with an OP37; this is simply because the bandwidth is now down to nearer 10 kHz because of the poor frequency response of the 741. The real contrast between the two devices will be apparent if the output from the circuit shown in Fig. 10.3 is taken to a simple audio amplifier and loudspeaker arrangement, and the noise is actually listened to. The OP37 produces a featureless hiss, the sound of true white noise. Every 741, in contrast, seems to have its own particular variety of noise. This subjective test is, of course, picking out the noise in the 1 kHz region, where the human ear is most sensitive. It is here that a device like the 741 may produce the most outrageous noise, or may be fairly quiet. It all depends upon the quality of the manufacturing process.

10.6 What happens to the input power?

With the circuit shown in Fig. 10.3 still in mind, it is interesting to consider a really classical problem of electronic circuit design and ask what really happens to the very small power input that such a circuit accepts. The dynamic microphone which would, in practice, be connected to the input, would produce a signal of, say, $10\,\mu V$ rms across the $90\,\Omega$ input impedance when the sound level was well above the circuit noise level. This is a power input of just over 1 pW. It might be thought that such a small quantity of power should be carefully led into the amplifying device, but this is not what happens at all. Virtually all of this input power is simply dissipated in the 15 kΩ resistor, R_1.

This is because the power gain of a device like the OP37 is so big, at frequencies in the audio range, that only a minute amount of power need be delivered to its true input. The function of the input circuit is only to get maximum power transfer from the signal source, the microphone, into the circuit itself. This power is then used to develop a voltage, which is the input signal for the amplifying device itself, and, at the same time, the optimum source impedance for this amplifying device must be arranged.

Things need to be more carefully done if low noise performance is called for at higher frequencies. As the signal frequency goes up, the power gain which may be obtained from an electronic amplifying device gets smaller and smaller until, when the microwave region is entered, it is absolutely essential to see that the device itself presents an impedance match to the signal source. Then all the input power should be usefully employed in operating the amplifying device itself. The idea of using a high input impedance amplifier, and then getting an input match by shunting this high impedance with a resistor, which, essentially, is what has been done in Fig. 10.3, is just not acceptable at high frequency.

Considerations of this kind come into many electronic instrumentation problems where a small high frequency signal must be dealt with, and the signal is so small that the amplifier must have the lowest possible noise figure. It is in this area that some very interesting circuit shapes are found. To illustrate this, the next few sections consider a problem which really belongs to the field of communications engineering: the repeater amplifier.

10.7 The repeater amplifier problem

A repeater amplifier must be designed so that it can be inserted into a cable, every 10 km or so, to make up for the losses in that cable and yet add as little extra noise as possible. The repeater amplifier, as shown in Fig. 10.4, should have an input impedance equal to whatever load impedance is put upon it. In practice, this would be the characteristic impedance of the cable: typically $50\,\Omega$ or $75\,\Omega$. Furthermore, the ideal repeater amplifier should present an output impedance equal to whatever impedance may be connected to its input. Only then can it be simply inserted into any cable.

Nordholt [14] has pointed out that the usual 'brute force' approach to the design of a repeater amplifier is first to build an amplifier, with the required gain and bandwidth, which has a high input impedance and a low output impedance. The correct terminating impedance for the cable, R_{in} in Fig. 10.4, is then obtained by putting an impedance in parallel with the high input impedance. Similarly, the correct driving impedance for the cable, R_L in Fig. 10.4, is obtained by adding an impedance in series with the very low input impedance of the amplifier. This approach is clearly not a good one, from a noise point of view, because of the power that must be lost in these added impedances. Exactly the same criticism came up in the previous section, where it was shown that virtually all the power input to the audio amplifier, shown in Fig. 10.3, was dissipated in the resistor R_1.

It may have been Norton [15] who first proposed some ways of getting around this problem, although the techniques which are really useful can be attributed to earlier work by Chaplin, Candy and Cole [16]. This earlier work will be considered in detail in section 10.10. Norton's proposals are particularly interesting, however, because they are examples of what has been presented here as the circuit shape approach to electronic circuit design. The first circuit proposed in Norton's paper is shown in Fig. 10.5.

In Fig. 10.5, an amplifier, A, is shown with negative feedback from output to input, arranged by means of two transformers, T_1 and T_2. It should be noted that the connection of these two transformers is very similar to that found in a directional coupler: the device considered in chapter 5 and shown in Fig. 5.12.

If the amplifier, A, in Fig. 10.5 has a high voltage gain, the voltage, v, across its input may be assumed negligible. Thus

$$v_{in} + v_{out}/N_1 = 0 \tag{10.5}$$

where N_1 is the turns ratio of T_1.

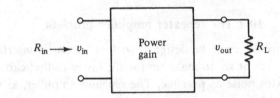

Fig. 10.4. *The repeater amplifier. Ideally, R_{in} is equal to R_L so that the amplifier may be inserted into a cable which has $Z_o = R_{in} = R_L$.*

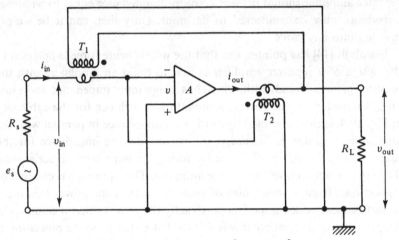

Fig. 10.5. *Feedback by means of two transformers.*

Similarly, if the amplifier, A, has a very high input impedance, its input current will be negligible, so that

$$i_{in} + i_{out}/N_2 = 0 \qquad (10.6)$$

where N_2 is the turns ratio of T_2. The amplifier system, as shown in Fig. 10.5, is thus an inverting amplifier and, if $N_1 = N_2 = N$, the voltage gain will be N, the input impedance will be equal to the load impedance, R_L, and the output impedance will be equal to the source impedance, R_s. The power gain will be N^2. All this is just what is required as a solution to the repeater amplifier problem.

The really interesting thing about Norton's proposal is the way in which the power input, $v_{in} i_{in}$ in Fig. 10.5, is nearly all passed on to the load, R_L, by means of the two transformers. The small amount of input power which is not passed on belongs to the terms which were neglected above, to obtain equations (10.5) and (10.6), and is the true input power of the amplifier A.

While very interesting as an idea, or a circuit shape, the amplifier shown

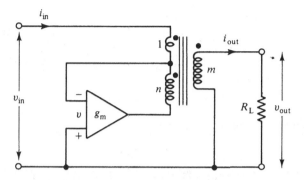

Fig. 10.6. *An OTA with transformer feedback.*

in Fig. 10.5 is very difficult to make in practice, at the high frequencies where the idea should really be applied, and has also been criticised by Nordholt [14] from a noise point of view. Norton [15] argued that, because feedback is provided by means of the lossless transformers, T_1 and T_2 in Fig. 10.5, the amplifier system is left with only the device noise. Nordholt [14] showed that this was not the case, although it is true that a better noise figure may be obtained with lossless feedback, compared to resistive feedback, and Norton [15] could demonstrate this experimentally.

Norton gave a second circuit in his paper [15] for a repeater amplifier which should avoid loss of power in its input termination. This circuit involved a single transistor in grounded base configuration, with negative feedback again provided through a transformer. As a circuit shape, this second circuit of Norton's suggests the one shown in Fig. 10.6, where an operational transconductance amplifier (OTA) is shown with transformer feedback. The transformer has a third winding to provide an output. This circuit leads to the next experimental circuit for this chapter.

10.8 An analysis of the circuit using an OTA

The OTA shown in Fig. 10.6 is a device which provides an output current, $g_m v$, where v is its true input voltage. A $1:n:m$ transformer is arranged to provide negative feedback and also a voltage output across a load, R_L. Note that 100 % negative feedback at d.c. is provided across the OTA when the signal source, v_{in}, is connected to the circuit through a capacitor. This circuit detail will be included in the next section. At the moment it is necessary to analyse just how the circuit shown in Fig. 10.6 is going to work.

Negative feedback at the signal frequency is provided by transformer windings '1' and 'n', because these windings are connected in the sense

shown by the spots in Fig. 10.6. There must be negligible current flow into
the inverting input of the OTA, however, so that the ampere–turn balance
in the transformer can only be provided when i_{out}, flowing in winding 'm',
balances the current i_{in} which flows in windings '1' and 'n'. This means
that

$$(1+n)i_{in} = (m)i_{out}. \tag{10.7}$$

The current i_{in} will be given by

$$i_{in} = g_m v \tag{10.8}$$

where v is the very small OTA input voltage. As i_{out} must be given by
v_{out}/R_L, where R_L is the load on the circuit shown in Fig. 10.6, it follows
that equations (10.7) and (10.8) may be combined to give

$$v = mv_{out}/g_m R_L(1+n). \tag{10.9}$$

The voltages across windings '1' and 'm' must be related so that

$$v_{in} - v = v_{out}/m. \tag{10.10}$$

Combining equations (10.9) and (10.10) gives the voltage gain of the
circuit as

$$v_{out}/v_{in} = m/[1 + m^2/g_m R_L(1+n)] \tag{10.11}$$

and this will equal m if the term $m^2/g_m R_L(1+n)$ is very small compared to
unity.

Norton's circuit, from which the circuit under discussion has been
derived, was intended to have an input impedance equal to the load
impedance R_L. When $v_{out}/v_{in} = m$ and $i_{out} = v_{out}/R_L$, equation (10.7) leads
to the result

$$R_{in} = v_{in}/i_{in} = (1+n)R_L/m^2 \tag{10.12}$$

so that the condition $R_{in} = R_L$ is obtained when

$$n = m^2 - 1. \tag{10.13}$$

Equation (10.13) shows that some difficulties will be found in applying
this circuit as a repeater amplifier at high frequency because, if a
reasonable gain is required, 20 db for example, the turns ratios, 1:n:m,
will be 1:99:10, and, at high frequency, this will not be easy. At audio
frequencies, such high turns ratios present no problems, but audio
amplifiers with the repeater property, $R_{in} = R_L$, are not in much demand.
From an experimentalist's point of view it is better to look upon the
circuit, shown in Fig. 10.6, as one which provides a voltage gain m and an
impedance transformation, given by equation (10.12), of $(1+n)/m^2$. These
properties are useful at audio frequencies, particularly when coupled with

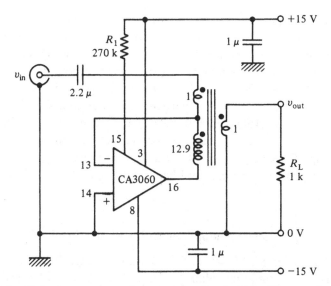

Fig. 10.7. *An experimental circuit using the CA3060. The transformer is an STC-66121B or RS-217-804 and is used with the two primary windings quite separate, to give input and output windings, and the two secondary windings in series to give the winding shown across the OTA.*

the low noise potential of the circuit that should follow from the lossless nature of its feedback elements. As in the previous circuit, Fig. 10.5, virtually all the input power to the circuit shown in Fig. 10.6 is passed on to the load, R_L, through the transformer.

10.9 An experimental circuit

A version of the circuit, first shown in Fig. 10.6, which can be built for experimental work is shown in Fig. 10.7. This uses the CA3060 OTA with its bias current, into pin 15, at just over 100 μA. This is set by R_1. The g_m of the CA3060 is about 0.1S, at this bias current, and is 3 db down at 100 kHz. This is ideal for audio applications.

The circuit shown in Fig. 10.7 uses the same type of audio transformer which was used in the first experimental circuit of this chapter, Fig. 10.3. This is connected to make $m = 1$ and $n = 12.9$, so that the resulting amplifier has unity voltage gain and a high input impedance. Furthermore, the condition $m^2/g_m R_L(1+n) \ll 1$, suggested by equation (10.11), is now very easy to meet, even when R_L is taken below the value of 1 kΩ shown in Fig. 10.7.

The voltage gain and input impedance of the experimental circuit should be measured and checked with the theory of the previous section.

The frequency response is determined by the transformer and should be flat across the 10 Hz–30 kHz band. It is interesting to reduce R_L and note how an optimum power output may be obtained when the OTA runs into both voltage and current limitations at the same peak output level.

The circuit shown in Fig. 10.7 has been put forward because it is unusual and is derived from a particularly interesting circuit shape, proposed by Norton [15] for high frequency and low noise work. The low frequency version given here has only limited practical application [17]. Norton's high frequency circuit, using a single transistor, gave a gain of 8 db over the band 5–200 MHz and had a noise figure of only 1.25 db at 100 MHz.

10.10 A repeater amplifier with resistive feedback

Looking back at Fig. 10.5, where feedback through the transformers, T_1 and T_2, caused the input impedance, v_{in}/i_{in}, to equal the load impedance, R_L, it is necessary to ask some more general questions about why this really happens. It turns out that the essential feature of this kind of feedback is that it involves a transfer of information about v_{in} and i_{in} to the *output*, and, at the same time, information about v_{out} and i_{out} to the *input*. In this way, changes in R_s cause changes in Z_{out}, while changes in R_L cause changes in Z_{in}: the very property required for a repeater amplifier.

This kind of multiple feedback can be done in many ways, and a solution using resistive feedback networks was published in 1959 by Chaplin *et al.* [16]. As a circuit shape, the idea can be put forward by considering the circuit shown in Fig. 10.8, where d.c. levels should be ignored for the moment and only the small signal changes, v_{in}, i_{in}, v_{out} and i_{out} considered. The analysis which follows is adapted from the very clear treatment given by Kovács [18].

Fig. 10.8 shows an amplifier in which the output stage, Q_1, is a series feedback circuit: the same circuit considered here in chapter 6, Fig. 6.3(*a*). This circuit, when working into a load, R_L, can provide a signal which is an accurate measure of the current in R_L because of the voltage which is developed across its feedback resistor, R_{E2}. As shown in Fig. 10.8, it is precisely this voltage across R_{E2} which is fed back to the input through resistor R_{F2}.

Similarly, feedback from output to input involving the value of v_{out} is fed back through R_{F1} into R_{E1}. This circuit consequently has the same feature as Fig. 10.5, discussed at the beginning of this section, as far as information about the output condition being fed back to the input. The

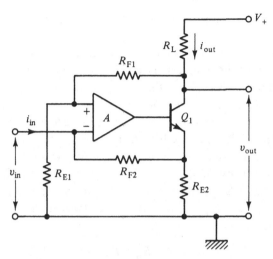

Fig. 10.8. *A first step towards a repeater amplifier using resistive feedback networks.*

following analysis will show that the same applies for input conditions influencing the output properties of the circuit.

If the gain block, A in Fig. 10.8, is ideal, having high input impedance and low output impedance, the open-loop voltage gain of the circuit shown in Fig. 10.8 will be simply AR_L/R_{E2}, all feedback being removed by open circuiting R_{F1} and R_{F2}. It may be assumed that AR_L/R_{E2} is very high. Obviously, restoration of feedback through R_{F2} will have no effect upon the gain v_{out}/v_{in}, because the components R_{F2} and R_{E2} can only change i_{in}, not v_{in}. Restoration of feedback through R_{F1} does fix the gain v_{out}/v_{in}, however, and this will be

$$v_{out}/v_{in} = (R_{F1} + R_{E1})/R_{E1}. \qquad (10.14)$$

The current flowing into the input terminals of the high gain block, A, will be negligible. This means that i_{in} must flow on through R_{F2}, and must be given by

$$i_{in} = [v_{in} + v_{out}(R_{E2}/R_L)]/R_{F2} \qquad (10.15)$$

so that, using equation (10.14) and writing $R_{in} = v_{in}/i_{in}$, it follows that

$$R_{in} = [R_{F2} R_{E1}/(R_L R_{E1} + R_{F1} R_{E2} + R_{E1} R_{E2})] R_L. \qquad (10.16)$$

Equation (10.16) shows that the repeater amplifier property, $R_{in} = R_L$, will be obtained if the designer can make $R_{F1} R_{E2}$ the dominant term in the denominator, and then also make the ratios $R_{F1}:R_{E1}$ and $R_{F2}:R_{E2}$ equal. This is the exact analogy of the identity, $N_1 = N_2$, between the turns ratios of the two transformers used for feedback components in the earlier circuit, Fig. 10.5.

To find the output impedance of the amplifier proposed by Fig. 10.8, a resistor, R_s, must be connected across the input terminals and R_L removed. A current, Δi_{out}, must then be injected into the output terminal and the resulting increase in output voltage, Δv_{out}, calculated.

The simplest way of looking at this calculation is to note that the change in the true input voltage to the high gain block, A, must be negligible. In the situation described above, the injection of Δi_{out} will cause the inverting input of A to rise by $\Delta i_{out} R_{E2} R_s/(R_{F2}+R_s)$, and the resulting increase in output voltage, Δv_{out}, will cause the non-inverting input of A to rise by $\Delta v_{out} R_{E1}/(R_{F1}+R_{E1})$. Equating these two increases then gives $R_{out} = \Delta v_{out}/\Delta i_{out}$ as

$$R_{out} = [R_{E2}(R_{F1}+R_{E1})/R_{E1}(R_{F2}+R_s)]\,R_s. \qquad (10.17)$$

Again, the repeater amplifier property, $R_{out} = R_s$, is obtained if the designer can make $R_{F1} R_{E2}$ and $R_{F2} R_{E1}$ the dominant terms, and also make $R_{F1}:R_{E1}$ equal $R_{F2}:R_{E2}$.

The above analysis confirms that the circuit shape shown in Fig. 10.8 is a possible basis for a good repeater amplifier design. Noise performance should be good because the terminating impedance for the input cable, R_{in}, and the driving impedance for the output cable, R_{out}, can be made to match the cable characteristic impedance, Z_o, but R_{in} and R_{out} are not found in the circuit as real resistors dissipating signal power. These essential terminating impedances have been determined by means of feedback resistors and, furthermore, the absolute value of these feedback resistors may now be chosen to optimise the active device source impedances from a noise point of view.

Clearly, the detailed design of high frequency amplifiers of this kind involves a great amount of calculation with accurate device modelling. The book by Maclean [19] deals with the design of this kind of amplifier and gives a complete treatment of the theory, with several examples of amplifiers which have been built and tested. The book includes photographs of the hardware and details of the computer aided design programs that were used. Another reference which gives hardware detail, along with the complete theoretical background needed, is the paper by Meyer, Eschenbach and Chin [20].

It is possible, however, to take quite a simple point of view in the design of an amplifier of this kind, provided it is to be used at fairly modest frequencies and intended to have quite high gain. High gain implies that only a small amount of feedback will be used, and this means that it will

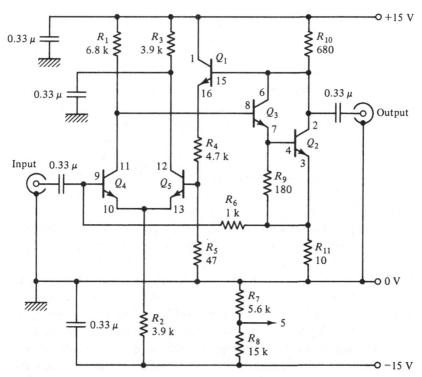

Fig. 10.9. An experimental repeater amplifier built around a CA3127E transistor array. Devices are numbered, and pin numbers given, according to the data sheet [21].

be easier to ensure amplifier stability. This is the approach adopted in the next section, where an experimental version of one of these interesting amplifiers is described.

10.11 An experimental repeater amplifier

The experimental wide-band repeater amplifier, shown in Fig. 10.9, is built around a CA3127E array of five high frequency transistors. These have a value of f_T in excess of 1 GHz, and are best operated with V_{CE} at 6 V. The collector currents should all lie between 1 mA and 10 mA to ensure high gain-bandwidth product.

Comparing the circuits shown in Figs. 10.8 and 10.9, it is clear that the high gain block, A in Fig. 10.8, is replaced by the long tailed pair, Q_4 and Q_5 in Fig. 10.9. This long tailed pair provides the inverting and non-inverting inputs that are needed for the two feedback paths shown in Fig. 10.8.

The series feedback output stage, shown as Q_1 in Fig. 10.8, is replaced by the series feedback output stage of Fig. 10.9, which involves the two transistors, Q_3 and Q_2. R_{E2}, in Fig. 10.8, is replaced by R_{11}, in Fig. 10.9, but R_L, in Fig. 10.8, is the true output load in Fig. 10.9, R_{10} being made high compared to R_L.

Feedback of the voltage across R_{11} to the inverting input is through R_6, in Fig. 10.9, so that R_6 plays the same role as R_{F2} in Fig. 10.8. Feedback of the output voltage, which is done through R_{F1} into R_{E1} in the prototype circuit shown in Fig. 10.8, is made through the emitter follower, Q_1 in Fig. 10.9, and then through R_4 into R_5. Introducing Q_1 avoids the problem of having to supply the d.c. called for by the feedback path, R_4 into R_5, from the output point.

The pin-out of the CA3127E makes access to Q_3 and Q_2 on one side of the 16 pin package, and access to Q_4 and Q_5 on the other side. For this reason, a choice of Q_4 and Q_5 as the input transistors, and Q_3 and Q_2 as the output transistors, makes the separation of output and input, in the layout of the circuit shown in Fig. 10.9, somewhat easier. Q_1 and Q_2 are the best matched pair of transistors in the CA3127E, but the choice of these as input devices leads to layout problems, and, in any case, good matching of the input devices is not called for in an *RC* coupled amplifier.

The resistor values shown in Fig. 10.9 are chosen to give a high open-loop gain to the circuit, and set sensible collector current levels and d.c. levels at the same time. High gain in the output stage means getting R_{11} down as low as possible, while high gain in the input stage means as high a value of R_1 as possible, taking into account the d.c. level called for at the base of Q_3, the collector current in Q_4, and the V_{CE} of Q_4. With the values chosen, I_{C4} is 2 mA and V_{CE4} is just under 2 V. For the other input transistor there is no problem about not getting a high value of V_{CE}: the problem is to keep V_{CE} down. The $V_{CE(max)}$ of the transistors in the CA3127E is 15 V, and this circuit is working from a ± 15 V supply.

Returning to equation (10.14), the gain of the experimental circuit, shown in Fig. 10.9, is set by the resistors R_4 and R_5, these playing the role of R_{F1} and R_{E1} shown in Fig. 10.8. As equations (10.16) and (10.17) showed, the repeater amplifier property of providing an R_{out} equal to the source impedance, and an R_{in} equal to the load impedance, is arranged by making the ratios of the feedback resistors, $R_{F1}:R_{E1}$ and $R_{F2}:R_{E2}$, equal. This means that, in the experimental circuit, $R_4:R_5$ should be made equal to $R_6:R_{11}$.

A gain of 100 has been chosen for the experimental circuit, which is only about one tenth of the open-loop gain so that there should be no problem with stability. Testing the amplifier with a sinusoidal input from a 50 Ω source, and with a 50 Ω load on the output, should confirm this gain of

40 db and show a 3 db drop in gain around 30 MHz, above which the gain should fall rapidly. This will be true up to output levels as high as 50 mV peak, where the output stage is having to supply ± 1 mA peak to the 50 Ω load. At output levels above this, some non-linearity will begin to be observed.

There are, of course, many other possibilities for an experimental circuit of this kind. Different types of transistor array may be used [22], discrete transistors with the same high frequency performance would allow a better layout, and an emitter follower buffer between Q_4 and Q_3, in Fig. 10.9, could be considered.

10.12 Measurement of input and output impedances

The most important feature of the experimental circuit shown in Fig. 10.9 is its ability to be inserted into a cable. The input impedance of the amplifier should equal its load impedance, and its output impedance should equal whatever source impedance is connected to it.

The input of the circuit shown in Fig. 10.9 works at very low level. The easiest way of checking that the input cable is correctly terminated by the amplifier input is to vary the length of cable in between the source, which should be a good quality 50 Ω or 75 Ω signal generator, and the amplifier itself. The output of the amplifier, measured across a good quality 50 Ω or 75 Ω termination, should be independent of the cable length. This test must, of course, be done at a high enough frequency. At 20 MHz, where the 30 MHz bandwidth amplifier might be expected to begin to show departure from the simple theory given in section 10.10, a $\lambda/4$ length of 50 Ω or 75 Ω cable is only a few metres long, and observing the apparent change in overall gain, as the length of the input cable is changed from $\lambda/4$ to $\lambda/2$, makes it possible to make a rough estimate of Z_{in}.

A more direct measurement of the output impedance is possible because the signal level at the output is high enough for direct observation. An interesting method is to use the directional bridge that was described in chapter 5, and built as the experimental circuit shown in Fig. 5.13. The output of the experimental amplifier, shown in Fig. 10.9, is connected to the load port, shown in Fig. 5.13, the source port is taken to a signal of 50 Ω source impedance, and the detector port is taken to a sensitive RF voltmeter which has a 50 Ω input impedance. This voltmeter will then show the level of mismatch at the amplifier output when the amplifier input has a good quality 50 Ω termination across it. This test is, of course, made with a sinusoidal source and the frequency is varied across the passband. The same information may be obtained if a pulse generator is used as a signal source and the RF voltmeter is replaced with a very sensitive wide bandwidth oscilloscope. It is very important, however, to keep the signal level at the output of the amplifier under test very low, at least below the 50 mV level at which non-linearity was found to appear in the gain measurements discussed in the previous section.

10.13 Noise measurements

The experimental circuit shown in Fig. 10.9 lends itself to simple noise figure measurements because the noise developed at the output is high enough to be observed with simple equipment. This is to be expected in view of equation (10.1). The rms noise voltage across the 50 Ω input termination for a bandwidth of 30 MHz at 290 K is just below 5 μV, so that a perfect amplifier with a gain of 100 would produce a noise output of 0.5 mV.

The experimental repeater amplifier will, of course, produce more output noise than 0.5 mV when its input is terminated with a good quality 50 Ω or 75 Ω termination, and its output is similarly terminated. Observing the output noise with a sensitive wide-band oscilloscope will show that accurate noise measurements may be far more difficult than the experimentalist expected. Unless the experimental circuit has been mounted inside a good quality metal box, the input and output connections are high quality co-axial connectors, and the power supply cable well decoupled at the point where it enters the circuit box, all kinds of unwanted noise may be observed. In urban environments one of the main problems will be the very high levels of television transmitter signals in the laboratory. While these are at frequencies more than ten times higher than the pass-band of the circuit under test, these television signals can be so large that they will still appear at the output. Such signals may be easily identified because their modulation is synchronised to the line frequency. Other radio signals which are a nuisance come from mobile communication equipment. These can be identified from the random nature of their coming and going.

Having overcome these environmental problems, the noise figure of the experimental circuit may be checked by using a standard noise generator test set. The usual technique is to increase the noise input until the noise power at the output is doubled [23]. Noise generators have improved greatly since the early instruments, which used thermionic diodes working under temperature limited emission. Modern noise generators use a high quality wide-band amplifier to amplify the thermal noise from a resistor at constant temperature, and this is followed by filters and attenuators to provide a test set noise output of variable bandwidth and level [24].

10.14 Noise reduction in special cases

The remaining sections of this chapter will consider some selected ideas that have been proposed to reduce noise in special cases. These are cases where either the thermal noise, the device noise, or some unwanted signal coming from the very nature of the signal processing system itself, can be reduced by using some special circuit. In the following three sections, one example of each of these three cases will be given.

10.15 Electronic cooling

Thermal noise is a problem in any very low level, wide-band, circuit that has to operate at room temperature. An obvious way of reducing thermal noise is to cool the circuit down well below room temperature, but this technique is only used in very special research environments. It is, for example, found in laboratories working in the field of experimental fundamental particle physics, where particle detectors are often used at very low temperatures to reduce the effects of thermal noise. This may be why the idea of 'electronic cooling' originated in the field of fundamental particle physics, although it has nothing to do with cryogenics, being a room temperature circuit technique, perhaps first described by Radeka [25].

The fundamental idea of electronic cooling is shown in Fig. 10.10. An amplifier is shown with negative feedback, provided by a single capacitor, C. At this initial stage, no questions are asked about the d.c. stability problems of this circuit idea.

Now suppose that the open-loop gain, $A(j\omega)$, of the amplifier shown in Fig. 10.10 has the form

$$A(j\omega) = A_0/(1 + j\omega/\omega_c) \tag{10.18}$$

which is, for example, typical of any operational amplifier having internal compensation. Well above ω_c the gain falls at a rate of 6 db/octave, and may be written

$$A(j\omega) \approx -jA_0(\omega_c/\omega). \tag{10.19}$$

The current fed back to the input terminal, i in Fig. 10.10, must be given by

$$i = -j\omega C v_{in} A(j\omega) \tag{10.20}$$

so that well above ω_c, where equation (10.19) applies

$$i = -\omega_c A_0 C v_{in}. \tag{10.21}$$

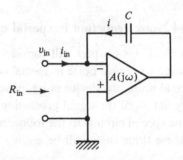

Fig. 10.10. An electronically cooled termination.

This current is independent of frequency because the increasing admittance of the capacitor, $j\omega C$, cancels out the falling gain of the amplifier, equation (10.19). It follows that, looking into the input terminals of the circuit shown in Fig. 10.10 a real positive resistance

$$R_{in} = 1/\omega_c A_o C \qquad (10.22)$$

will be seen.

Now it can be argued that this resistor, R_{in}, is 'noiseless' if the amplifier, $A(j\omega)$ shown in Fig. 10.10, is also noiseless. The reasoning is that the feedback through C involves no dissipation. When the amplifier, $A(j\omega)$, does have some intrinsic noise, a return must be made to Fig. 10.1 where the noise of an amplifier is represented by two generators, e_n and i_n. For any practical value of R_{in} the contribution from i_n may be neglected. This means that the resistor, R_{in}, is left looking as though it is in series with a noise voltage generator e_n, the e_n belonging to the amplifier, $A(j\omega)$, in Fig. 10.10. This e_n may be much lower than the $(4kTR_{in})^{\frac{1}{2}}$ which would represent the noise of a real resistor, R_{in}, over unity bandwidth.

From a noise point of view, then, R_{in} looks like a resistor which is not at room temperature. Suppose a practical version of the circuit shown in Fig. 10.10 was made using the LT1028 operational amplifier, which, as Table 10.1 shows, has $e_n = 0.85$ nV/Hz$^{\frac{1}{2}}$. At *room* temperature, 290 K, this is the noise voltage expected from a resistor with a value of 45 Ω. It follows, by simple proportion, that R_{in} *appears* to be at a temperature of $290(45/R_{in})$. So, if a value of C is chosen to make R_{in}, given by equation (10.22), equal to 450 Ω, then this 450 Ω would appear to be 'electronically cooled' down to 29 K: well below liquid nitrogen.

What value of C would, in fact, be needed to make $R_{in} = 450 \Omega$? The LT1028 has a gain–bandwidth product of 50 MHz ($A_o \approx 7 \times 10^6$ and $\omega_c/2\pi \approx 7$ Hz). Equation (10.22) shows that $C = 7$ pF would give $R_{in} = 450 \Omega$. This is rather a small value of capacitance in practice, and the problem of providing d.c. feedback across the LT1028, without spoiling the whole project, is going to be difficult. Nevertheless, this discussion shows that the circuit idea shown in Fig. 10.10 is a valuable one.

Electronic cooling has been applied by Gatti, Manfredi and Marioli [26] to provide a low noise termination for a transformer coupled radiation detector, and they give references to earlier work. The idea is interesting, very similar in its philosophy to the ideas of Norton [15], discussed in section 10.7, and subject to the same criticisms given by Nordholt [14] concerning dissipationless feedback in general.

10.16 Using devices in parallel

A circuit idea which attempts to reduce intrinsic device noise is worth considering briefly because it may be an example of a circuit designer doing the right thing for the wrong reasons. This is the idea that several transistors connected in parallel will give a better noise figure than a single device. In the author's experience, this idea is assumed to be correct by quite a few people, although it appears to have been explicitly published only once [27].

The reasoning behind the idea that several amplifiers in parallel are better than one, is that the noise from each amplifier is random, there is no correlation between the noise from one amplifier and its neighbour, but that the signal output from each amplifier is the same and these outputs may be summed. If there are n amplifiers, perhaps the signal to noise ratio at the output will be improved by $n^{\frac{1}{2}}$, because the signal will have been amplified by n while the noise will have been amplified by only $n^{\frac{1}{2}}$.

This will not happen because it is the noise *power* which matters, not its instantaneous value expressed as a voltage or a current. Looking back at equation (10.4) will confirm this: it is the product $e_n i_n$ which decides the noise figure. Nevertheless, better noise performance may be achieved by using several transistors in parallel for two reasons.

The first reason is that a parallel connection of bipolar transistors gives a composite device with a lower base spreading resistance, usually referred to as $r_{bb'}$ [28, 29]. This resistance contributes to $\overline{e_n^2}$ by an amount $(4kTr_{bb'})\ V^2/Hz$. In fact, high frequency, low noise, bipolar transistors, of the highest quality, are actually made, internally, as a parallel connection of several individual devices. This has been the case for a very long time [30]. The composite transistor has interdigited base and emitter contacts which reduce $r_{bb'}$ to a minimum.

The second reason why a parallel combination of bipolar transistors may lead to a lower noise figure is that the designer may then be able to

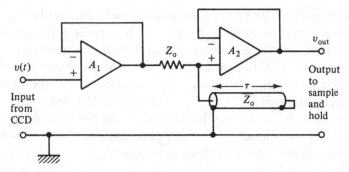

Fig. 10.11. *The reflection delayed noise suppression technique due to Ohbo et al.* [31].

get the input conditions such that the circuit is working closer to the source impedance that gives minimum noise figure: the $R_{s(opt)}$ given above by equation (10.3). A better design would, of course, be made by choosing the correct single device.

10.17. An example from high definition television

Finally, it is interesting to consider an example of reducing an unwanted random signal which turns up in some signal processing system because of the very nature of that system. The example chosen concerns a high definition, solid state, colour television camera.

Completely solid state television cameras use an array of photodiodes which deliver their outputs to charge coupled devices (CCDs) arranged vertically and horizontally, these CCDs giving a serial data output.

Flicker noise and noise from surface states in a solid state device of this kind are considerable. The video output will have a large random low frequency component, and this problem is usually overcome by clamping the output just before the signal level from one pixel is about to arrive at the output, and then measuring only the change in level. This method is very difficult to apply at the very high data rates involved in high definition work. A camera described by Ohbo, Akiyama and Tanaka [31], having 2×10^6 pixels, produced its output on two interlaced channels working at over 37 MHz.

The solution described by Ohbo *et al.* [31] is a beautiful example of a circuit shape or circuit idea. It is shown in Fig. 10.11. The video signal from the CCD is buffered by amplifier A_1, which must provide a very low output impedance. This low output impedance is then made up, by means of the resistor Z_o in Fig. 10.11, to match the characteristic impedance of a short circuited length of delay line.

The input of the delay line is connected to the input of the second amplifier, A_2 in Fig. 10.11. This amplifier is a high input impedance voltage follower. It follows that the input voltage to amplifier A_2 is the sum of two signals. The first is the video signal at time t, divided by two because it is arriving at the input of A_2 from the potential divider made up from resistor Z_0 loaded by the delay line. The second signal is the *inverted* video signal, delayed by 2τ and again divided by 2. This second signal is inverted because of its reflection at the short circuit, delayed by twice the length of the line, and halved because the reflected signal sees the correct termination, Z_0, when it arrives back at the input of A_2.

It follows that, if the video signal is $v(t)$, the output from the circuit shown in Fig. 10.11 is

$$v_{out} = [v(t) - v(t - 2\tau)]/2. \qquad (10.23)$$

If this signal is now made the input to a sample and hold circuit, which samples only during the instant when the CCD is giving the pixel output, then all the low frequency flicker noise and surface state noise from the solid state camera output signal will be removed. Only the *change* in output signal will be sampled and held, this change being precisely the pixel intensity that is needed to make up the true video signal of the scene being televised.

10.18 Conclusions

In the field of electronic instrumentation the circuit designer might bear in mind a remark made by Faulkner, in a paper which was cited above [28], 'we must avoid the assumption that noise considerations are a sort of "extra" which only needs to be taken into account under exceptional circumstances'. In other words, good circuits are low noise circuits. From this point of view, the EMC considerations, mentioned at the beginning of this chapter, may be the most important. Electronic instruments must work alongside other pieces of equipment, and this means that these instruments must be designed so that their circuits do not pick up unwanted signals. For ideas in this area, the books by Wilmshurst [1] and Ott [2] are invaluable.

Notes

1 Wilmshurst, T. H., *Signal Recovery from Noise in Electronic Instrumentation*, Adam Hilger, Bristol, 1985, p. ix.
2 Ott, H. W., *Noise Reduction Techniques in Electronic Systems*, John Wiley, New York, second edition, 1988.
3 Gray, P. R., and Meyer, R. G., *Analysis and Design of Analog Integrated Circuits*, John Wiley, New York, second edition 1984.
4 Bell, D. A., *Noise and the Solid State*, Pentech Press, London, 1985.

5 Buckingham, M. J., *Noise in Electronic Devices and Systems*, Ellis Horwood, Chichester, 1983.

6 Morse, P. M., *Vibration and Sound*, McGraw-Hill, New York, 1948, p. 227. Fig. 52 on this page shows the threshold of hearing to be well below 10^{-16} W/cm^2 for frequencies between 2 kHz and 5 kHz. The human ear has an aperture of about 1 cm^2.

7 Note 3 above, pp. 695–6. A particularly clear treatment is also found in the book by M. H. Jones: *A Practical Introduction to Electronic Circuits*, Cambridge University Press, Cambridge, second edition, 1985, pp. 44–7.

8 Widlar, R. J., *Proc. Nat. Electronics Conf.*, **21**, 85–9, 1965.

9 Erdi, G., Schwartz, T., Bernardi, S., and Jung, W., *Electronic Engineering*, **53**, No. 652, 57–71, 1981.

10 *Linear Databook Supplement 1988*, Linear Technology Corp., Milpitas, California, pp. S2.21–S2.36.

11 There is a valuable discussion of the relative merits of the bipolar transistor and the JFET, from a noise point of view, in P. Horowitz and W. Hill, *Art of Electronics*, Cambridge University Press, Cambridge, second edition, 1989, pp. 441–5.

12 Note 4 above, pp. 74–7.

13 Note 3 above, pp. 667–76.

14 Nordholt, E. H., *IEEE Trans. Circ. Syst.*, **CAS-28**, 203–11, 1981.

15 Norton, D. E., *Microwave Journal*, **19**, No. 5, 53–5, May 1976.

16 Chaplin, G. B. B., Candy, C. J. N., and Cole, A. J., *Proc. IEE*, **106B**, Suppls. 15–18, 762–72, May 1959.

17 O'Dell, T. H., *New Electronics* **20**, No. 2, 20 January, 1987.

18 Kovács, F., *High-frequency applications of semiconductor devices*, Elsevier, Amsterdam, 1981, p. 140 *et seq.*

19 Maclean, D. J. H., *Broadband Feedback Amplifiers*, John Wiley, New York, 1982.

20 Meyer, R. G., Eschenbach, R., and Chin, R., *IEEE J. Sol. St. Circ.*, **SC-9**, 167–75, 1974.

21 RCA Data Bulletin File No. 662.

22 O'Dell, T. H., *Electronic Engineering*, **60**, No. 744, 21–2, 1988.

23 Horowitz, P., and Hill, W., *Art of Electronics*, Cambridge University Press, Cambridge, second edition, 1989, p. 450.

24 Rasaratnam, D. K., *Hewlett–Packard J.*, **38**, No. 7, 30–6, July 1987.

25 Radeka, V., *IEEE Trans. Nuc. Sci.*, **NS-21**, 51–64, 1974.

26 Gatti, E., Manfredi, P. F., and Marioli, D., Nuc. Inst. Meth., **193**, 539–47, 1982.

27 Grocock, J. A., *Wireless World*, **81**, 117–18, 1975.

28 Faulkner, E. A., *Radio and Electronic Engineer*, **36**, 17–30, 1968.

29 Sutcliffe, H., *Int. J. Elect. Eng. Ed.*, **6**, 371–4, 1968. Sutcliffe also commented on note 27 in *Wireless World*, **81**, 264, 1975.

30 Lee, H. C., *Microwave Journal*, **12**, No. 2, 51–65, Feb. 1969.

31 Ohbo, M., Akiyama, I., and Tanaka, T., *IEEE Trans. Con. Electr.*, **CE-35**, 368–74, 1989.

Name index

213

Subject index